Praise for *The Five Forces That Change Everything*

"Steve Hoffman's book, *The Five Forces*, is a compelling vision of our forceful futures (of connectivity, biotech, space, automation & intelligence), and a great guide to what you can do to prepare. Steve is in a rare position to weigh in on these diverse topics as investor/advisor to so many rising entrepreneurs. You will be shocked how close and fast these asteroids of progress appear to be."
—**Dr. George Church**, professor of genetics at Harvard Medical School, and professor of health sciences technology at the Massachusetts Institute of Technology (MIT) and Harvard

"*The Five Forces* poses complex and profound questions about how technology is shaping our world and what will be the future of the human race. AI and robotics can bring enormous benefits for societies but can also be used for malicious purposes if falling into the wrong hands. Therefore, we must be prepared to keep pace and use it responsibly, without compromising the fundamental principles of Human Rights."
—**Dr. Irakli Beridze**, head of the Centre for Artificial Intelligence and Robotics, United Nations Interregional Crime and Justice Research Institute

"A lucid and fascinating exposé of the myriad technologies and scientific breakthroughs that are already shaping our lives, and which will determine what kind of people we become in the next few decades. Utterly fascinating and essential easy reading."
—**Dr. David Levy**, international chess master and author of more than 40 books on chess, AI, and computers

"Anyone who cares at least one ounce about the future must read this book. Wake up. Hoffman offers an entirely new perspective on what is about to happen in the world."
—**Dr. Sebastian Thrun**, cofounder of Udacity, CEO of Kitty Hawk, and former director of the Stanford Artificial Intelligence Lab

"Steve Hoffman nails it! He identifies the science fiction that is becoming our new reality. A delightful run through a wild future of sentient AIs, cyborgs, transgenic animals, and designer babies. Read it!"

—**Tim Draper**, founder of DFJ, included in the *Forbes* "Midas List of Top Venture Capitalists," AlwaysOn #1 Most Networked Venture Capitalist, and *Worth* magazine's "100 Most Powerful People in Finance"

"An exciting whistle-stop tour of how technology will affect our future. This is right on the button in terms of accuracy and coverage. A brilliant book!"

—**Dr. Kevin Warwick**, professor of cybernetics at Coventry University

"Hoffman steps on the gas on page one and doesn't ease off until the last word. It's common knowledge how fast science and technology are accelerating, but this book forces you to feel the g-forces involved. Strap in and enjoy the ride."

—**Dr. Josh Bongard**, Veinott Professor of Computer Science, University of Vermont

"In his trademark lively style, Steve Hoffman shines a light on the key technologies most likely to impact the future of humanity. If you want a knowledgeable tour guide to your future, hop on board and buckle up!"

—**Dr. Jerry Kaplan**, serial entrepreneur, lecturer at Stanford University, and author of *Artificial Intelligence: What Everyone Needs to Know*

"*The Five Forces* framework delivers an incredible itinerary of insights into our evolving future. It should be required reading for all innovators. Hoffman offers a unique invitation to discover how today's cutting-edge innovations drive our tomorrow."

—**Dr. Aimee Arnoldussen**, innovation and commercialization specialist at Discovery to Product, UW Madison

"Thought provoking and insightful: Steve manages to provide a great overview of the technological disruptions happening in the world. *The Five Forces* is a great source for new ideas and business disruption."

—**Dr. Ott Velsberg**, chief data officer at the Ministry of Economic Affairs, Estonia

"A highly entertaining and informative book that explores the forces shaping our future and how these will transform our lives."

—Dr. Steve Horvath, professor of human genetics at UCLA

"Steve Hoffman's book is the reminder that the future of humanity relies on the groundbreaking technology we have now. It is on us to decide what kind of world we want the future generations to live in."

—Dr. Alysson R. Muotri, director of the Stem Cell Program at the Institute for Genomic Medicine, UC San Diego

"Change is constant and our evolution to adapt will define quality of life. *The Five Forces* takes an intriguing look into the rapid evolution of technology that connects humans more intimately with technology, from technology inspired by biology to those that mimic biology. Understanding and appreciating the interplay between the five forces will set readers thinking, imagining, and innovating."

—Dr. Sharath Sriram, professor of engineering at RMIT University and chairman of Policy Committee of Science & Technology Australia

"*The Five Forces* is very thought provoking. It is an incredible glimpse into the future that suggests that emerging technologies could go very well or go very badly."

—Gary Wishnatzki, cofounder of Harvest CROO Robotics

"*The Five Forces* provides a great summary of the most recent advances in AI, with a very focused and objective, yet creative, view of where we stand today. It outlines a futuristic, but also brutally honest, stance of where the five fundamental forces may drive humanity in the years to come."

—Dr. Oggi Rudovic, machine learning researcher at MIT Media Lab

"Hoffman is that rare combination of the insider who brings the fierce curiosity of an Outsider's mindset to new worlds. Read *The Five Forces* to open your own thinking to the astounding developments that are already here, and around the corner!"

—Jonathan Littman, coauthor of the international bestsellers *The Art of Innovation* and *Ten Faces of Innovation*

"As a venture capitalist active in the United States and Asia, Steve Hoffman sees start-ups pitching him today what the future will look like tomorrow and shares this vision with us in this compelling book."

—**Andrew Romans**, general partner at 7BC Venture Capital and author of *The Entrepreneurial Bible to Venture Capital*

"The questions raised in this book are probably the most important ones to deal with today, for the sake of our future. It is not about agreeing with Steve but allowing yourself to be taken on a full cycle of thinking so you can eventually make up your mind. Do it for yourself."

—**Liviu Babitz**, cofounder and CEO of Cyborg Nest

"Hoffman's writing is at once brilliant and lucid, hitting home at a time in society when technology is transforming the world and impacting our lives more than ever."

—**Dr. Steve Mann**, professor of electrical and computer engineering at University of Toronto, cocreator of the discipline of Humanistic Intelligence, along with Marvin Minsky and Ray Kurzweil, and cofounder of InteraXon and BlueberryX

THE FIVE FORCES
THAT CHANGE EVERYTHING

Also by Steven S. Hoffman

Make Elephants Fly (2017)

Surviving a Startup (2021)

THE FIVE FORCES
THAT CHANGE EVERYTHING

HOW TECHNOLOGY IS SHAPING OUR FUTURE

Steven S. Hoffman

Matt Holt Books
An Imprint of BenBella Books, Inc.
Dallas, TX

The Five Forces That Change Everything copyright © 2021 by LavaMind LLC

All rights reserved. No part of this book may be used or reproduced in any manner whatsoever without written permission of the publisher, except in the case of brief quotations embodied in critical articles or reviews.

Matt Holt Books is an imprint of BenBella Books, Inc.
10440 N. Central Expressway
Suite 800
Dallas, TX 75231
benbellabooks.com
Send feedback to feedback@benbellabooks.com

BenBella is a federally registered trademark.
Matt Holt and logo are trademarks of BenBella Books.

Printed in the United States of America
10 9 8 7 6 5 4 3 2 1

Library of Congress Control Number: 2021005936
ISBN 9781953295040
eISBN 9781953295422

Copyediting by Miki Alexandra Caputo
Proofreading by Lisa Story and Denise Pangia
Indexing by Amy Murphy
Text design and composition by Yara Abuata
Cover design by Sarah Avinger
Cover image © Shutterstock / antishock
Printed by Lake Book Manufacturing

Special discounts for bulk sales are available.
Please contact bulkorders@benbellabooks.com.

Dedicated to Mike, Sharna, Doug, Zachary, and Skylar Hoffman.

Special thanks to Naomi Kokubo and Randy Ludensky.

CONTENTS

Foreword by Dr. Steve Mann · xv

The Forces · xix

FORCE 1: MASS CONNECTIVITY | 1

The Neural Pioneers	4
Neural Implants and Brain Chips	8
Next-Gen Brain-Machine Interfaces	12
Mind Assistants and Brain Applications	15
Brain Hacking, Mind Control, and Mental Privacy	16
The Next Wave of Virtual Reality	21
Artificial Sensory Perception	24
Is the World a Simulation?	28
Hyperlayers: Augmented Reality	30
Mixed Reality: Living a Multimodal Existence	33
The Seventh Wave of Mass Connectivity	35

FORCE 2: BIO CONVERGENCE | 37

Wetware Warriors: Biohackers on the Loose	38
Grinders and Cyborgs	40

Reshaping the Human Anatomy	44
Smart Drugs, Energy Boosters, and Super Supplements	47
Life Extension: The Epic of Gilgamesh	51
The Global Impact of Life Extension	58
Cryogenics, Rebirth, and Rabbit Brains	59
The Clones Are Coming	62
The Bionic Body	66
The CRISPR Revolution	69
Gene Drives	73
Cow Cultures: Lab-Grown Meats	76
DNA and Organic Computers	78
Chimeras, Bioprinting, and Transgenic Life-Forms	80
Genetic Imagineers: A Brave New World	85

FORCE 3: HUMAN EXPANSIONISM | 95

Quantum Computing	96
The Birth of New Materials	99
Nanoscale: Manipulating Molecules and Atoms	102
Nanobots: Miniature Medical Machines	106
Masters of the Universe or Mass Extinction?	110
Mission to Mars	113
Building an Interplanetary Ecosystem	116
Space Pioneers and the Gold Rush	120
The Space Economy	123
Warp Drive: Far-Out Ideas	125
Life on Other Planets	128

FORCE 4: DEEP AUTOMATION | 133

Smart Cities: Dreams of Magic Kingdoms	135
Smart Governments	138
Robo Cops: Autonomous Policing	140
The State of Surveillance: AI Is Watching You	143
AI Fortune-Tellers: The Prediction Machines	149
Reinventing Education	154
Algorithmic Art: Can AI Be Creative?	159
Silicon Valley and the Future of Hollywood	164
Lights-Out Factories and Supply Chain Automation	174
Dr. Roboto: Smart Hospitals and Health Tech	177
Algorithmic Agriculture	180
Robots to the Rescue	185
The Future of Work: A Jobless Society?	191

FORCE 5: INTELLIGENCE EXPLOSION | 195

Singularity: The Superintelligence Is Coming	197
Can Machines Be Conscious?	199
Society and Sentient Machines	202
Humanoid Robots and Emotional Machines	205
Can You Love a Robot?	211
AI Bosses: Working for the Bot	218
Self-Learning AIs and Evolutionary Robots	220
AI Economies: The Centralization of Power	225
Superintelligence: A Philosopher King in a Box	228

Hatching Demons: Will AI Destroy Us?   230

Merging with Our Machines   235

Brain Net: Tapping the Subconscious   239

Uploading Ourselves: Hyperconnected Realities   242

Augmenting Our Senses: Neuroscience and Perception   247

Organoids and Biological Super Brains   253

The Coming of Super Sentience   256

The Culmination of Forces   259

Notes   263

Index   283

About the Author   297

FOREWORD

Dr. Steve Mann

Steve Hoffman's writing is at once brilliant and lucid, hitting home at a time in society when technology is transforming the world and impacting our lives more than ever. The accelerated pace of change is at times difficult for society to comprehend; however, by beginning with the brain as the epicenter of Mass Connectivity, this book gives us a deep understanding of the human condition.

Hoffman outlines five forces, which I can summarize in the following five words: brain, cyborg, quantum, automation, and singularity.

Force 1 is the brain and its associated connectivity. I've personally witnessed this force through co-founding InteraXon, which makes the Muse brain-sensing headband, as well as co-founding BlueberryX, makers of the brain-and-world-sensing spectacles that monitor our physical, affective, and mental health constantly, even underwater. As our brains become connected in a massively parallel wearable computer network, the social fabric of our society is shifting toward the inclusion of affective computing/health, as well as new ways of communicating and interacting with both the digital and physical worlds.

Force 2 is the cyborg age. The cyborg age is the post-human age, also known as transhumanism or post-humanism, in which the very nature of what it means to be human is challenged. I feel a personal sense of connection to this force, as I've often been referred to in the media as "the world's first cyborg," and although I don't like the term personally, it has caught on in mainstream culture. In my childhood in the 1960s and 1970s, I experimented with adapting and modifying myself in many crazy ways. Some of these were childhood stupidity and some resulted in new inventions that had widespread acceptance in the world. This includes wearable computing and wearable artificial intelligence (AI), which I brought to MIT when I founded the Wearable Computing Project. While at MIT, back in 1991, Charles Wyckoff and I coined the term eXtended Reality, which became X-Reality and XR intelligence. Today, we can see all these pieces coming together, laying the foundation for the cyborg age.

Force 3 is the quantum age of nanoscale and calls to mind what I might call the post-post-human age or the post-cyborg age. This is where we'll be able to develop new materials and nanomachines that not only give us superhuman powers but fundamentally transform the material world around us.

Force 4 is the age of automation, AI, and machine intelligence. It is not only an important force on its own but represents a force multiplier when combined with the other forces. Endowing machines with human-like intelligence feeds back directly into the cyborg age and will amplify and accelerate the technological transformation already underway.

Force 5 is the singularity or intelligence explosion. At MIT, we used to say that when an equation contains a singularity, you'll get a division-by-zero error, meaning some quantity will blow up. In a sense, what we're seeing is division-by-zero on a massive scale. Software replicates itself at near-zero cost and in near-zero time. The result is something Marvin Minsky, Ray Kurzweil, and I called the "sensory-singularity"—that is when sensory intelligence explodes over near-zero time and cost. This ultimate force will be the equivalent to splitting atoms in its power and potential to transform and even destroy our species.

All of these ideas come together in *The Five Forces That Change Everything* in a thoughtful and accessible way, so that anyone trying to stay on top of the latest trends in technology can gain a sense of where we're headed and how these fundamental breakthroughs will alter our lives and the world around us.

Dr. Steve Mann has been recognized as "the father of wearable computing" (IEEE ISSCC 2000), and "the father of wearable augmented reality (AR)" for his invention of the "Digital Eye Glass"

(EyeTap) and mediated reality, a predecessor of AR. He also invented the Chirplet Transform, Comparametric Equations, and High Dynamic Range (HDR). Mann received his PhD from MIT in 1997 and is a professor of electrical and computer engineering at the University of Toronto and cofounder of InteraXon and BlueberryX, as well as the co-creator of the discipline of Humanistic Intelligence, along with Marvin Minsky and Ray Kurzweil.

THE FORCES

Let me introduce myself. My name is Steve Hoffman, but in Silicon Valley they call me Captain Hoff because I'm the captain and chairman of Founders Space, one of the world's leading start-up accelerators. I'm also a venture capitalist, and I've spent the past several years traveling the globe, collaborating with top entrepreneurs, scientists, and visionary thinkers. This has given me unparalleled access to the research and people developing the technologies that will fundamentally impact every person on the planet.

I'm about to take you on an extraordinary journey into the minds and ideas of the people poised to reshape our world. I will reveal how new scientific breakthroughs and business ventures are primed to transform our lives and turn science fiction into fact. From Silicon Valley biohackers and lifelike robots in Japan to the world's first three-parent baby in the Ukraine, I will give you an inside look at the limits of what's possible and the incredible impact these developments will have.

Each step of the way, I'll show you how these discoveries are part of the five fundamental forces driving humanity forward at an unprecedented pace. We're living at a time when the impossible is becoming all too real, and because of that we must understand where our imagination and ingenuity are taking us. Our creations not only have the power to enable humans to live longer, healthier, richer lives but also the potential to permanently change and even destroy us. The decisions we make in the coming years will determine who and what the human race becomes.

As our machines become capable of doing most of the work people have performed for centuries, we are headed for a massive social reorganization. For example, nanotech is at the point where it's possible to unleash invisible robots and new materials into the world that can forever alter life on Earth. With genetic engineering, we have taken evolution into our own hands, creating new species of plants and animals that never existed before. We're developing AIs that can predict future events, create lifelike simulations, and learn from their own mistakes. And space technology has advanced to the point where colonizing other planets and discovering alien life-forms is within our reach.

As the impetus behind the technological changes we see happening around us, the five forces will determine the direction humanity takes. Understanding them is crucial because their reach extends across our social structures, from how our economies develop and our institutions function to what will become of our species. To illustrate this, the book is arranged into five sections:

1. **Mass Connectivity**, the force driving humans to plug our machines and our brains into interconnected, intelligent, digital networks, will radically transform how we live and work, fusing the material and virtual into a new, alternate reality.
2. **Bio Convergence**, the force driving humans to merge biology with technology, will enable us to decode the building blocks of life, create entirely new species of plants and animals, conquer disease, and heighten human abilities.
3. **Human Expansionism**, the force driving human beings to push to the edges of our known universe, will propel us further into the quantum world and deeper into outer space, in order to harness their vast potential.
4. **Deep Automation**, the force driving humans to algorithmically automate all the underlying processes for managing, growing, and sustaining life, will accelerate innovation, create wealth, and free us from the need to work.
5. **Intelligence Explosion**, the force driving us to develop new forms of superintelligence that far surpass human capabilities, will bring into existence sentient machines that run our economies, act as our champions, and merge with our consciousness.

For each of the forces, we will delve into the hard questions. How will we adapt to a world with billions of intelligent devices watching, analyzing, and responding to our every move? What would it feel like to connect our brains directly to the internet and interact with one another? Should we modify the genetic code of life to produce new crops, cures for cancer, and DNA-edited babies? What does it mean to merge with our machines and establish a class of cyborgs? And what happens when AI reaches or exceeds the level of human intelligence?

As we tackle these questions, I will expose you to numerous scientific discoveries and illuminate the overarching changes underway. My goal is to ignite your imagination, as I reveal what lies ahead. Join me on this extraordinary journey and find out where the five forces will lead us.

FORCE 1

MASS CONNECTIVITY

Mass Connectivity, the force driving humans to plug our machines and our brains into interconnected, intelligent, digital networks, will radically transform how we live and work, fusing the material and virtual into a new, alternate reality.

For *Homo sapiens* to move beyond tribes and build a civilization required a communications revolution. This came in the form of the written word. Writing has permitted people to connect with one another over great physical distances and spans of time. It has allowed us to pass down our accumulated knowledge, traditions, and beliefs from generation to generation. Without writing, we could not have expanded trade, organized religions, and built advanced societies. In other words, innovation and progress would have stagnated without written language.

The next major communications breakthrough came with the invention of Gutenberg's printing press. This accelerated the exchange of ideas between people and allowed for the mass dissemination of knowledge. It precipitated the Renaissance, the Enlightenment, and, ultimately, the scientific revolution. The greater the flow of information, the more rapidly civilization progressed.

With each key technological milestone, from the pony express to the telegraph to the telephone, our ability to communicate with one another and share knowledge has increased exponentially. This leads us to modern times and the advent of computers. What makes these machines so powerful is not just their ever-improving processing power, storage capacity, and portability, but also their ability to talk with one another.

Connecting computers together to form the internet has unleashed the explosion in innovation we see around us. You can compare the internet to the pathways that connect regions of the human brain. There are billions of neurons in every human brain. Each neuron can have thousands of synapses. Without the trillions of synaptic connections, our brains wouldn't function. In the same way, we have built a global network composed of billions of human brains and computers working together. That's what makes the internet so powerful.

Since the ARPANET (the precursor to the internet) went live in the late 1960s, each significant network upgrade has spurred unprecedented innovation and productivity gains across the economy. The transition from narrowband to broadband and wired to wireless gave birth to countless new products and services. Even the smallest upgrades matter because our networks sit at the heart of our innovation ecosystem, and every improvement affects everyone and everything connected to that network.

For entrepreneurs, investors, researchers, and corporations to innovate, they need to collaborate, exchange information, and access the latest technologies. Otherwise, they won't be first to market with the next generation of products and services. Upgrading the communications infrastructure won't change the world in and of itself, but some entrepreneur taking advantage of this network will create the next Microsoft or Google.

What we have to remember is that technologies don't exist in isolation. They are all part of a single system. Sometimes, you can have the most advanced piece of technology ever conceived, but without the ecosystem to support it, it's nothing but an academic experiment. There is no clearer example of this than machine-learning algorithms. They had been around for nearly half a century, but their real utility could not be realized until there was a critical mass of people interacting online. Without mass connectivity, these algorithms could not tap into the vast amounts of high-quality data necessary to create products like AI assistants, logistics and supply chain automation, and autonomous vehicles.

The same is true for YouTube and Netflix. There were dozens of online video services before the advent of accessible high-bandwidth networks, but they all failed. It was only with the coming of broadband that they took off. Without advances in mass connectivity, they wouldn't exist. So, what comes next? Where are we headed from here? Is there another major leap forward in connectivity coming our way?

Yes, and it may even rival the internet itself in terms of magnitude. This next step will be when we successfully develop a robust, high-bandwidth connection from our brains to the internet. By translating our thoughts directly into digital commands, while simultaneously enabling our brains to access massive amounts of cloud-based information and processing power, we will forever change not only how we communicate but how we perceive and interact with the world. And when you combine brain-computer interfaces (BCIs) with other technologies, like virtual and augmented reality devices, there will be an explosion in new innovations.

Today's VR and AR headsets and primitive brain-reading devices only scratch the surface of what's possible. In the coming decades, we'll look back on these devices much like we do early punch-card computers. As mass connectivity moves away from the limits of physical interfaces and toward direct mental interactions, our entire concept of the internet will change. We won't access the applications: We will live inside them.

Imagine what it will be like to walk around with a chip inside your head that allows you to tap into vast amounts of information, process this data, communicate brain to brain with other people, and even exchange fully formed concepts and memories. Will we come to think of the digital and physical as one and the same? Will virtual objects look, behave, and even feel as real as anything in the material world? What will it be like to interact with lifelike avatars and experience a fully augmented reality? These are just a few of the questions to ask as we dive into the next wave of mass connectivity.

On this journey, you'll learn about the people and research behind some of the most advanced brain-computer interfaces, neural prosthetics, and cognitive platforms. I'll also expose you to breakthroughs in dream recording and artificial sensory perception, while posing hard questions, like: What limits should we place on mind-reading devices? Is it possible to hack a brain and rewrite memories? What are the ethical concerns around having technology that can access our innermost thoughts? And what will it be like to live and work in a mixed-reality world?

THE NEURAL PIONEERS

Machines that can analyze our brainwaves and record our thoughts are being developed in laboratories and companies around the world. These are called brain-computer interfaces, and they have a long history. In 1924, Hans Berger, a German neuroscientist, invented the first electroencephalogram (EEG), which is not that different from those widely used today.

Berger's fascination with brainwaves stems from an incident he had as a young man. Initially, he enrolled in university as a mathematics student with the goal of becoming an astronomer; however, after only one semester, he dropped out and joined the German cavalry. One day, he was out riding his horse and had an accident. Luckily, he wasn't seriously injured, but for some reason, his sister, who was many kilometers away, felt something terrible had happened and made his father telegram him.

Berger later wrote that he believed "It was a case of spontaneous telepathy, in which at a time of mortal danger, and as I contemplated certain death, I transmitted my thoughts, while my sister, who was particularly close to me, acted as the receiver."[1]

After this, Berger became obsessed with figuring out how his mind could have transmitted a signal to his sister. He returned to university determined to study medicine, in the hope of discovering the physiological basis of this "psychic energy." That eventually led him to develop the first EEG device, which could detect brainwaves.

An EEG is simply an array of sensors placed on a person's head that measures changes in the electrical activity in the brain. The EEG can sense voltage changes in ionic current within and between neurons. Let's be clear, an EEG cannot read your thoughts, but it can record the electrical signals emanating from your brain and a computer can match patterns. If a computer knows that a specific pattern is associated with a certain mental state, word, or action, it can decipher this information and understand something about your thought processes.

Today's start-ups are using EEG devices to capture digitized signals and send them to computers for storage and data processing. Analyzing the resulting brainwave patterns can tell us a bit about what is going on inside the brain. For example, beta waves are associated with a person being awake, attentive, and alert, while alpha waves are associated with being relaxed, calm, and lucid. Other waves are associated with different mental states, like that aha moment when someone recognizes something. Although the waves are low resolution, entrepreneurs are using EEGs to create all sorts of applications, from mobile apps that help people meditate to software for tracking mental fitness.

The advantages of EEG devices are that they're noninvasive—meaning you don't have to insert probes inside your brain—and relatively inexpensive. You can go out and buy a brain-sensing headband for a few hundred dollars or less online. However, the problem with the current consumer-grade EEG devices is that they often don't work that well. The signals tend to be noisy because the connections between the electrodes and the scalp are imperfect. There are often too few electrodes in consumer devices to get accurate readings. Hair also gets in the way and any movement of the skin can generate false signals. That said, in the lab, EEGs can work pretty well. Keep in mind, we're still in the early days of brain-computer interfaces, and things will only improve. Some of the biggest changes are coming in the form of more sophisticated AI that can filter out the noise and better decipher complex brainwaves.

Neurable, a Boston start-up, is using machine learning to measure and classify EEG signals in order to provide greater accuracy. Ramses Alcaide, who showed his entrepreneurial spirit at a young age, is the visionary CEO behind the company. He was four years old when his father moved the family from Mexico to the United States in search of better opportunities. When he was six, he started to buy and fix broken video games, then sell them for a profit. By the age of nine he was earning money repairing computers. After getting his PhD in neuroscience from the University of

Michigan, he launched Neurable, which perfectly combined his intellectual interests and knack for business.

When we met, Alcaide shared his plans to build a brain-computer operating system. Much like we have iOS and Android for smartphones, he believes we will need an advanced operating system for the next generation of brain-computer interfaces. Neurable is taking the first step toward this by developing a pair of brain-controlled consumer headphones that can control music directly through your thoughts.

Founded by CEO Spencer Gerrol, an innovative academic with a background in cognitive psychology, SPARK Neuro is another fascinating neurotech start-up. Its technology analyzes people's brainwaves as they watch video content. Using the biofeedback data, the company helps TV networks to produce better shows and advertisers sell more products. SPARK Neuro's product functions like a focus group, except that instead of asking people for their opinions, the EEG device captures their immediate neural activity, and a deep-learning algorithm determines what they actually think. This is important because what people say they think during a focus group and what they truly feel are often different. Many times, participants will conform to group opinion and say what they think other people want to hear. SPARK Neuro's brain-computer interface and deep-learning algorithm reveal how we actually feel at a fundamental, nonverbal level.

So, how do they do it? SPARK Neuro demonstrates that when people are watching playful videos of babies, puppies, and kittens, their system can clearly detect the changes in brainwaves. It mirrors the changes seen when people are happy, which may explain why those super cute animal videos go viral. They even tested the technology on independent voters leading up to the 2016 presidential elections. The AI predicted that Donald Trump would win based on the reactions of the people participating.

Can this type of brain science lead advertisers to sell more products? Some of the world's biggest brands believe so. SPARK Neuro's clients include Hulu, NBC, Barclays, GM, and Anheuser-Busch. It's possible that if enough consumers adopt brain-computer interfaces for everyday use, gathering this type of data may become as common as analytics on websites.

Halfway around the world in South Korea, SOSO is a start-up that uses EEGs in education and health care. Over seemingly endless cups of soju and plates of spicy food, we discussed the future of brain-computer interfaces. Dongbin Min, the CEO, explained how this technology can help students learn to focus better. As students play a series of learning games that run on tablets, they wear EEG devices that monitor their brainwaves and provide neurofeedback. The games use the incoming data to alter the gameplay in real time in order to improve each student's performance and concentration levels. It can be particularly valuable to children with learning disorders. Min's team is using similar neurofeedback games to enable elderly people to improve memory function and slow cognitive decline. This field is still in its infancy, but Min is determined to figure out how brain-computer interfaces can be used in novel ways to advance both education and health care.

On the cutting edge of EEG research, Adrian Nestor, a leading neuroscientist and professor at the University of Toronto, has conducted experiments with far-reaching implications. He had test subjects look at an image on a computer screen. Then, using an EEG cap, along with machine-learning algorithms, he was able to digitally reconstruct the image using only the signals from the brain. In other words, he managed to pluck an image right out of someone's head and transfer it to a computer.

There are many possible uses for this technology. In the future, you may be able to take your visual memories and upload them as images to the cloud. Or, if you wear an EEG cap at night, you may be able to capture images directly from your dreams. Another possibility is to enable people who are unable to verbally communicate, like stroke victims, to express themselves with mental images instead of words.

"It could also have forensic uses for law enforcement in gathering eyewitness information on potential suspects, rather than relying on verbal descriptions provided to a sketch artist," says Nestor.[2] Someday lawyers may use images captured from the brains of witnesses as evidence in a court of law.

If that's not wild enough, researchers at the University of California, Berkeley, ran an experiment where they captured video images of what test subjects were visualizing in their heads. To accomplish this, they had to go beyond EEG and use functional magnetic resonance imaging (fMRI), which measures brain activity by detecting changes associated with blood flow.

In an experiment, volunteers laid down inside an fMRI machine for hours while watching Hollywood movie trailers. The brain activity was fed into a computer program that learned to associate visual patterns in the movie with corresponding brain activity. The final results were a blurry but recognizable video reproduction of what the volunteers were seeing inside their heads.

At this stage, it isn't ready for primetime. No one wants to sleep inside a bulky fMRI machine every night simply to record their dreams. However, eventually, technology like this may allow us to wake up in the morning and watch our dreams.

Despite promising results in labs around the world, the current generation of devices are still limited. If we really want to jump straight into the future of mass connectivity and link our brains to the internet, nothing beats implanting a brain chip. Would you mind letting someone pry open your skull and insert some sensors? This technology is headed our way, like it or not.

NEURAL IMPLANTS AND BRAIN CHIPS

If you look at the evolution of *Homo sapiens*, our brains haven't changed in the past 30,000 years. We have essentially the same brains that our prehistoric ancestors had when they were living in caves. However, the world around us has changed dramatically. How is it possible that our prehistoric brains can function in a modern, high-tech society that's profoundly different from the thousands of years we spent as hunter gatherers? It's because our brains are incredibly malleable.

The human brain is not hardcoded. It has a high degree of plasticity. Its neural pathways can be easily overwritten. New habits and behaviors rewire our brains throughout our lifetimes. Every time you learn a new skill—like how to play a musical instrument or sport—the neural circuits in your brain change. This allows humans to do everything from writing poetry to building skyscrapers. It also enables us to adapt to vastly different environments.

It was essential for our prehistoric ancestors, who had to learn to live in places as diverse as the Amazon rainforests, the Kalahari Desert, and the Andes mountains. The human brain's flexibility has enabled us to dominate the Earth and reshape the planet to suit our needs. Yet, in spite of all the progress we've made, we still know relatively little about how our brains actually function. We don't even have a full understanding of our own consciousness.

Using brain-computer interfaces, scientists are just beginning to unravel the mysteries that lie beneath our skulls. Miguel Nicolelis, a pudgy, persuasive Brazilian with a charming habit of waving his arms when he speaks, is one of the leading researchers in this field. As a professor of neurobiology, he has spent the past three decades designing a series of experiments that have pushed the limits of brain-computer interfaces.

As a child, Nicolelis became fascinated with science during long afternoons of backyard explorations with his grandmother. Upon graduating with a doctorate from the University of São Paulo, he went to his adviser, César Timo-Iaria, the father of Brazilian neuroscience, and said that he was tired of listening to one neuron at a time. At this rate, it would take a billion years to complete his research. He wanted to record hundreds of neurons simultaneously and listen to a brain symphony. Timo-Iaria told him to catch a flight to the United States and find someone crazy enough to fund the research.

Nicolelis took this advice and eventually landed at Duke University, where he began to push the boundaries of his field. In 2002 his team took a rhesus monkey named Aurora, opened up her skull, and implanted an array of ninety-six tiny electrodes. These electrodes connected her brain to a robotic arm that controlled a joystick in a video game. Using just her thoughts, she was able to figure out how to move a cursor across the screen. Every time she successfully moved the cursor over a target, she received a reward of orange juice.

In the next experiment, they took a monkey, referred to simply as Monkey K, and increased the number of microfilament arrays inserted into its brain (6 arrays of 96 microfilaments each, for a total of 576 microelectrodes) and connected them to a computer that controlled the movements of a wheelchair. The monkey quickly learned to control the wheelchair just by thinking and became highly proficient at navigating toward a plate of succulent grapes.

Nicolelis realized that if a monkey can control robotic arms and wheelchairs with its brain, humans could surely do the same. In his next experiment, he decided to try something different. This time he connected the brains of two rats directly to the internet. They were placed in separate cages in different cities. When a tasty treat was available, scientists turned on a yellow light in the first rat's cage. The rat learned to push a lever to obtain the treat whenever the light was showing. The second rat received no indication that a treat was available. But because its brain was directly connected to the first rat over the internet, it did receive brain signals. Within a short time, the second rat learned to interpret the first rat's brain signals and press the lever. "These experiments showed that we have established a sophisticated, direct communication linkage between brains," says Nicolelis.[3]

In essence, Nicolelis had successfully transferred thoughts from one living brain to another through the internet. If you think about this, it's an astonishing achievement. It means that we have the capability to transfer information directly between our brains. All we have to do is implant electrodes and hook ourselves up to the internet.

Nicolelis went on to tweak the experiment so that whenever the second rat successfully decoded the brain signals from the first rat and obtained a treat, the first rat received an extra reward. The rats' brains began to subconsciously synchronize to maximize the benefits. This feedback loop prompted the first rat to send clearer signals, improving communications between the rats and increasing the number of rewards. The rats didn't even realize that their brains were working as a single unit.

Today, rats and monkeys aren't the only ones getting brain implants; humans are volunteering, too. At BrainGate, a collaboration between Stanford and Brown, researchers have embedded chips the size of a baby aspirin into the brains of quadriplegics, enabling them to control a robotic arm with their thoughts. "One of the participants told us at the beginning of the trial that one of the things she really wanted to do was play music again," says Paul Nuyujukian, a bioengineer at Stanford. "So, to see her play on a digital keyboard was fantastic."[4]

The quadriplegics were also able to interact with applications on a computer tablet using their thoughts alone. "The tablet became second nature to me, very intuitive," says one of the testers. "It felt more natural than the times I remember using a mouse."

How liberating it must be for anyone who has lost the use of their arms or legs to suddenly be able to use a computer again, play music, feed themselves, and drive about in a wheelchair. And this is only the beginning. After an unfortunate accident, Dennis Degray was paralyzed from the collarbones down, but he's now able to text message his friends with the help of a brain chip.

At the University of California, San Francisco, scientists have managed to turn brain signals into speech with help from an AI. They had volunteers with electrodes implanted in their brains train the AI to recognize their brain signals and output a synthetic approximation of the participant's voice. "For the first time, this study demonstrates that we can generate entire spoken sentences based on an individual's brain activity," says Edward Chang, a professor of neurological surgery.[5]

Elon Musk believes this is the future. His start-up Neuralink has raised hundreds of millions of dollars to commercialize and build upon the research done at these universities. Neuralink wants to improve the lives of people with neurological conditions, like Alzheimer's and Parkinson's, as well as those suffering from brain damage and spinal cord injuries. However, its ultimate goal is to create a tertiary level of the brain that will be linked to artificial intelligence.

"With a high-bandwidth brain interface, I think we can have the option of merging with AI," says Musk.[6]

To get there, Neuralink wants to make implanting a brain chip as painless and simple as LASIK surgery. They envision an outpatient surgery that requires only local anesthesia, where lasers drill tiny holes in your head, and the chip slides right in.

Despite all the progress being made, I don't recommend getting a brain chip anytime soon. There's still a long way to go. Electrodes can dissolve and corrode inside the body, and no one knows the life span of these devices or the long-term effects of living with them. Even if the technical hurdles can be overcome, the biggest issue holding back mass adoption is people's fear of embedding a foreign object in their brains.

NEXT-GEN BRAIN-MACHINE INTERFACES

MIT researchers have taken an entirely different approach. They have come up with a device called AlterEgo that doesn't read brainwaves at all but instead relies on a process called subvocalization. When you say words to yourself in your head, you activate muscles around your vocal cords. AlterEgo can sense these tiny muscle movements and decode them. The result is silent speech.

"Our idea was: Could we have a computing platform that's more internal, that melds human and machine in some ways and that feels like an internal extension of our own cognition?" says Arnav Kapur, who began working on the project as a graduate student at MIT.[7] "Imagine perfectly memorizing things, crunching numbers as fast as computers do, silently texting other people, suddenly becoming multilingual so you can hear the translation in your head in one language and speak in another."[8]

Kapur grew up in New Delhi, India, and comes to the project with his own perspective. He wants to build technologies that augment us, not replace us. His goal is to design devices that stimulate our curiosity and creativity while boosting our cognitive abilities. Kapur's own inventiveness seems to know no bounds. Even though he's still in his twenties, he's already invented a 3D-printable drone, experimented with measuring gene expressions at large scale, developed an audio device that narrates the world for the visually impaired, and collaborated on designing a lunar rover.

The team at MIT has gotten AlterEgo to the point where it can recognize basic keywords like "What time is it?" and reply through a bone-conduction headset with answers the user can hear. It can also perform rudimentary tasks, like adding up numbers or moving a cursor on a screen. Probably the best thing about AlterEgo is that it is not a true brain-computer interface, which means it can't read your deepest thoughts. It can only interpret the words you subvocalize.

"We believe that it's absolutely important that an everyday interface does not invade a user's private thoughts," says Kapur. "It doesn't have any physical access to the user's brain activity. We think a person should have absolute control over what information to convey to a person or a computer."9

For all the progress they've made, the MIT prototype is still a work in progress. When I contacted Kapur to ask if it was ready to be commercialized, he told me it was still some ways off. Given time and money, however, his goal is to make the device "invisible," meaning it will be less distracting than a pair of wireless earbuds.

AlterEgo is just one of many new technologies coming into existence right now that has the potential to change the future of brain-computer interfaces. Another is epidural electronics, also called e-tattoos. E-tattoos are thinner than a sheet of paper and as flexible as a Band-Aid. You can simply apply them to your skin like stickers, and they can begin to read your brainwaves. European researchers are testing e-tattoo electrodes that are as accurate as traditional EEG devices and can be produced cheaply using an inkjet printer. It's only a matter of time before start-ups begin marketing brain-computer interfaces that you can stick behind your ear or under your bangs, and no one will be the wiser.

There are also other technologies coming that could make brain reading far more accurate than EEG. Quasiballistic photons, for example, have the ability to penetrate the skull and see what's happening inside the brain, and they are far more accurate than EEG devices. Excited by the potential, back in 2017 Mark Zuckerberg formed a special hardware division called Building 8 to research applications for such technologies. There was a huge amount of hype when Facebook made the initial announcement, but not much has materialized in the years since. That's because this method of extracting neural information is still too slow for a consumer device.

Other technologies that are being tested around the world include ultrasound, radio frequencies, magnetic fields, and electric fields. Some researchers are even exploring nanotransducers, which are tiny particles the width of a human hair that can transform external magnetic energy into an electric signal inside the brain. They could act as a brain-computer interface when injected into someone's head. Another possibility is to use viruses to insert DNA into cells, altering the cell function so that they act like nanotransducers.

Theodore Berger, a biomedical engineer at the University of Southern California, is a thinker, tinkerer, and dreamer. He grew up in a household that placed a premium on introspection and accomplishment—the type of accomplishment that has an impact on the people around you. Realizing that psychology doesn't have the tools to fully understand the cause and effect of behavior, Berger has dedicated his life to figuring out this puzzle. He's working on developing an artificial hippocampus to transform short-term memories into long-term ones.

"We're not putting individual memories back into the brain," says Berger. "We're putting in the capacity to generate memories."[10] This is the first step in learning how the brain stores long-term memories. If successful, the next step could be figuring out how to write new memories directly to the brain.

In fact, the Defense Advanced Research Projects Agency (DARPA) is investing tens of millions of dollars in researching smart helmets and other bidirectional devices with the intended goal of reading and writing data to soldiers' brains. They're calling this project Next-Generation Nonsurgical Neurotechnology. "By creating a more accessible brain-machine interface that doesn't require surgery to use, DARPA could deliver tools that allow mission commanders to remain meaningfully involved in dynamic operations that unfold at rapid speed," says Al Emondi, an expert on neurotechnology and human-machine interaction at DARPA.[11]

DARPA's Carnegie Mellon University team is working on a completely noninvasive device that will use ultrasound waves to guide light into and out of the brain to record neural activity, while using interfering electrical fields to write to specific neurons. DARPA's Rice University team aims to develop a minutely invasive, bidirectional system for reading from and writing to the brain. For the reading function, the device will use diffuse optical tomography. For the writing function, the device will use a magneto-genetic approach to make neurons sensitive to magnetic fields.

Apparently, the US Department of Defense believes it's possible to both read and write to the brain. The value to the army is clear. In a chaotic, noisy battle scenario, soldiers using brain-computer interfaces could share information and collaborate with cloud-based AI, satellites, drones, tanks, and a variety of robots in real time. This could turn the entire fighting force, both humans and machines, into a single potent weapon.

So, where does that leave us? It's difficult to say because technological progress isn't linear. It comes in jerky spasms. A real breakthrough could materialize tomorrow, or it may be years or even decades away. What we do know is that it will happen. There's no doubt that brain-computer interfaces will one day be capable of connecting us seamlessly to the internet.

MIND ASSISTANTS AND BRAIN APPLICATIONS

How will brain-computer interfaces change our daily lives? Will these devices eventually replace the smartphone as our primary communications tool? There's a good chance they will, especially if they are designed to work seamlessly with our cognitive functions, so that interacting with a mind assistant feels as natural as talking to ourselves. After all, we are continually having a dialogue with ourselves in our heads. Why shouldn't we invite an advanced AI to join these conversations, especially if it can help us solve our problems, feel better, and perform at a higher level?

This sophisticated AI would be designed to integrate into our cognitive flow, so that it feels effortless and natural—not forced. Unlike talking to Siri, Alexa, or Google Home, which can be a bit tiresome, a well-designed cognitive AI would merge with the flow of our thoughts, so that we don't realize it's there until we need it. It would continually monitor our internal dialogue, anticipate our needs, and be ready to respond to whatever it determines we want. If we are struggling to recall a fact, without being prompted, it would activate an intelligent agent to search the internet for that information and present it to us. Sending a message, making a call, or striking up a conversation with someone could be as simple as thinking of the person's name and then dashing off a message in our minds.

At the same time, we could browse the internet in our heads. Whatever we wanted to know would only be a thought away. There would be no reason to memorize information when we could just look it up online in real time. If we wanted to save something for quick retrieval, we could enhance our brains with virtually unlimited cloud storage. We may eventually be able to upload and save any images formed in our brain or send these to our friends or coworkers, much like we share photos and videos today.

We may also be able to tap into the processing power of computers around the globe to help us perform complex calculations. To further augment our intelligence, there could be a host of specialized brain applications. Just like we download apps for our smartphones, we could select from a wide array of mind apps that would enable us to organize activities, share experiences, play games, access stock markets, translate foreign languages, plan out our travel schedules, and more.

These apps would be designed to work naturally with our underlying cognitive processes. Their goal would be to make us far more effective, while simplifying our lives. They may perform tasks like reminding us of appointments, scheduling meetings, managing work-related projects, looking after our finances, and following up on countless chores. We may even tell our mind assistants to act on our behalf, negotiating business deals, signing legal contracts, buying gifts for relatives, and keeping an eye on our kids.

At some point, we may come to look at our mind assistants as extensions of ourselves. We may come to trust them to act in our best interests. Instead of worrying about something, we may delegate many of our problems to our mind assistants and let them come up with the best solutions. But, as miraculous and useful as this may sound, it could come at a price. Giving anyone access to our brains, especially large corporations, poses serious privacy and security risks. What if we wind up getting hacked?

BRAIN HACKING, MIND CONTROL, AND MENTAL PRIVACY

Even if brain-computer interfaces could dramatically improve the quality of our lives, allow us to perform better on the job, and open up incredible new experiences, do we really want these devices accessing our innermost thoughts?

Granted, most of us put up with big companies, like Facebook, Amazon, Apple, and Google monitoring everything we do online, from chatting with friends and browsing the web to what apps we use, which products we buy, and where we are at any given time of day. You would think more people would care about their privacy, but they don't. Most people just can't be bothered. The convenience of using a smartphone outweighs most concerns they have over their data.

Companies like Facebook have taken advantage of this complacency in order to maximize their revenue. Facebook is now facing challenges from governments around the world for abusing people's data. Many countries, including the European Union, have enacted stricter privacy laws, and some politicians in the United States want to see Facebook broken apart or strictly regulated.

At the same time, the average person has essentially surrendered. People have not left Facebook in droves, even after the company lied to them repeatedly, handed over their data to disreputable sources, and showed little respect for their privacy. Facebook's stock has rebounded since the depths of the crisis, and although Facebook paid lip service to the complaints, it's clear they're still prioritizing profits over privacy.

If governments don't step in and protect people's privacy, will data derived from brain-computer interfaces be any more secure than our online data? It's doubtful. If governments are lax on regulating what these devices can do, corporations will naturally take advantage of this to grow their businesses. The majority of companies will not voluntarily forfeit the right to use brain-derived data to market products, sell ads, deliver personalized services, or predict what people will do and how they will act.

So, we must ask ourselves if we really want these devices, and do we trust the government to protect us? Even if the government imposes regulations, will the corporations developing the hardware and software respect those rules? And if they don't, what punishments will be imposed?

Once we begin walking around with our brains connected to the internet, how can we be sure our mental privacy will be respected? It's one thing for us to knowingly put something out there on the internet, understanding full well that companies like Facebook are tracking our every action. But how will we know for certain which of our thoughts a BCI device is monitoring and how far it will go with this information?

Let's consider some hypothetical scenarios. Say you are at work and have a crush on one of your coworkers. Maybe this person is your employee, and you begin to fantasize. You may be able to control your actions, but it's much harder to control your thoughts. If everyone uses brain-computer interfaces at work to improve their productivity, what does this mean for you? Would your private thoughts now be logged on company servers? Could you be fired for mental sexual harassment? Maybe not, but what if you later make a mildly off-color remark, and your coworker lodges a complaint? What happens to this data if you enter a court of law?

Even if nothing ever comes of the data, do you really want a company to have access to the most intimate parts of your psyche? Would you ever feel comfortable putting on a brain-computer interface that could potentially scour your brain for thoughts, emotions, and memories and use them to make money, or worse?

That said, if you're online right now, you are probably being tracked, and with enough data, a smart algorithm can begin to analyze your psyche. It can even build a fairly accurate personality profile. So, is there a difference?

This is why corporations will bend over backward to assure us that these devices are safe. Facebook already stated that its future brain-computer interface would be similar to MIT's AlterEgo device, in that it would only record what users subvocalize. But can we trust any company with something as personal as our thoughts?

I believe there needs to be an independent way to secure our brain data. This may come in the form of a future blockchain or some other highly secure storage system, where we retain control and the rights to all of our brain data. Maybe we implant special hardware to limit access to parts of our brain. For every technical problem, there is a technical solution, but at some level, all of these require a degree of trust because someone or something always has to create the hardware and software.

So, is there any way we can guarantee that corporations will safeguard our data? Probably not. Remember, people were surprised to find out that actual humans were listening in on their Alexa conversations. These listeners were Amazon employees doing their jobs, but isn't it still disturbing? Many of us have an Alexa in our bedrooms, and we don't always monitor what we are saying or doing when these devices are turned on.

Shouldn't we be dubious of any promises the big tech companies will make and even more concerned with how governments may use our brain data? Edward Snowden exposed how the National Security Agency was conducting secret mass surveillance programs on people all over the world. They did this without any public oversight and outside the limits of the US Constitution. Would there be any reason to think they might not develop a backdoor to commercial brain-computer interfaces? And it's not just the US government. Most other countries wouldn't hesitate to do the same thing.

I can see a world where authoritarian governments may even compel their populace to get brain implants, so they can better control their actions. This dystopian future is all too real when you consider how some governments today are monitoring social media, filtering internet content, and even installing spying apps on smartphones. It's just a matter of time before some country requires all its citizens to wear brain-computer interfaces at all times.

Even scarier, what happens when someone uses this technology not just to monitor people's brains but to rewrite their memories and control their actions? No, this isn't a bad sci-fi movie. Researchers at Zhejiang University have actually performed mind control on rats using a brain-computer interface. "The results showed that rat cyborgs could be smoothly and successfully navigated by the human mind to complete a navigation task in a complex maze," wrote the researchers.[12]

In this experiment, whenever the human manipulator thought about moving his left arm, the rat with the brain chip was commanded to turn left. Whenever he thought about moving his right arm, the rat turned right. Whenever he blinked, the rat moved forward. The entire time, the rat behaved as if it was making the decisions itself.

At the University of Texas Southwestern Medical Center, researchers implanted false memories in the brains of birds that changed how they sing. "We identified a pathway in the brain that if we activate, it can implant false memories for the duration of the syllables," says Todd Roberts, whose team used lasers to manipulate activity at the synapses between neurons.[13]

The thought that a hacker may get into your brain is horrifying. If your brain is connected to the internet, no amount of security can guarantee your safety. Anything can be hacked. Right now, it's bad enough if crooks hack your smartphone and steal your identity. They may be able to ruin your credit and make your life miserable for six months. But if you are wearing a brain-computer interface, identity theft takes on a whole new meaning.

Hackers may not be content to steal your money and passwords; they may want to steal your past, writing over your memories and reprogramming your mind. You may become their slave without ever realizing it.

Are we headed down the road to a cyberpunk, dystopian, Orwellian nightmare where thought police can monitor our every move, hackers can implant false memories and control us, and our most private thoughts belong to someone else? I hope not. But how do we address this? The evil genie is about to come out of the bottle.

Even if the world chooses to ban the use of brain-computer interfaces, like we did with chemical weapons, would that be enough? Eventually someone will develop a brain-computer interface powerful enough to read our thoughts at a distance using lasers, ultrasound, or other technology. Instead of passing through a metal detector at an airport, people may be forced to pass through a brain scanner. If we wind up in a world like this, it may turn all of us into paranoid schizophrenics who run around with tinfoil on our heads.

Elon Musk has already stated that his goal with Neuralink is nothing short of merging "biological intelligence with machine intelligence." He is worried about saving humanity from an all-powerful AI, but who is going to save us from ourselves? Throughout history, humans have tended to be our own worst enemies.

It's unlikely that most governments will force people to adopt brain-computer interfaces in the near future. But what if we do this voluntarily? All we have to do is allow the natural forces of capitalism to take over, and we may come to find that these devices are so useful and compelling that we will gladly sacrifice even our innermost thoughts for the advantages they provide.

This may sound far-fetched but look at our smartphones. Most of us can't live without them, even for a few hours. We need them with us all the time, and we seldom consider the data we're giving away. True brain-computer interfaces may become even more essential to our lives. There may well come a point where we cannot compete or even function in the world without the assistance of a brain-computer interface. A person without a brain-computer interface may come to be treated like a subhuman, whose lack of augmented intelligence and ability to communicate brain-to-brain are simply unacceptable in the workplace or social settings.

This is a scary but all too real possibility. Remember, people are social animals. Most of us will do whatever our peers are doing. Humans want to be part of the group. It's in our DNA. We are more afraid of being left out than being manipulated. That's the story of human history, and history has a way of repeating itself.

So, what will actually happen in the future? Will we wind up creating heaven on earth, where human beings can enhance their brainpower, collaborate better, and reach a higher plane of existence, or a hell, where we wind up as soulless zombie slaves? Most likely neither extreme will come to be. Instead, connecting our brains to the internet will probably turn out much like the internet itself. There will be positives and negatives in roughly equal proportion.

We will be able to connect with people in ways we never imagined, improve our mental capabilities and productivity, and create a richer, more dynamic world, but in the process, we may lose the privacy of our own thoughts and a degree of control over our lives. Just as we are at the mercy of our smartphones now, we will find our lives irrevocably changed, and we will have to adapt to our new environment, as we have done so many times before.

THE NEXT WAVE OF VIRTUAL REALITY

When considering the next stage of mass connectivity, brain-computer interfaces are just one piece of a future system that will also include virtual and augmented reality. People will eventually use brain-computer interfaces to interact with one another in entirely virtual environments.

Though we've had a taste of virtual reality (VR) already, we are still in the early days of VR. It's like comparing 1970s arcade games to modern video games. When virtual worlds reach maturity, they will become indistinguishable from our reality. They'll include everything from visual and audio to touch, taste, and smell. They will be part of the fabric of mass connectivity, which will change how we communicate, work, play, and socialize with one another.

To understand how far we've come, let's step back to the beginning of virtual reality. VR has gone through many permutations, dating all the way back to the Victorian era. The first virtual reality device was developed in 1838 by Charles Wheatstone, a British inventor and physicist. It was called the stereoscope and worked by using twin mirrors to project a single image, creating the illusion of three dimensions. It's the same concept as one of those simple, plastic toys you can hold up to your eyes with the slides that look somewhat 3D. This was a clever device, but it took another 150 years and many failed experiments before virtual reality came into its own.

The next milestone came in 1956, when Morton Heilig, a Hollywood cinematographer, invented the Sensorama, an ambitious project that resembled a 1980s-era video arcade game. He wanted to create a multi-sensory machine that would take people inside a movie. The Sensorama offered the experience of riding a motorcycle on the streets of Brooklyn. Users felt the vibration of the motorcycle seat and simulated wind on their faces, while being immersed in a 3D view of the city, complete with smells wafting past. Unfortunately, the Sensorama proved too expensive to turn a profit.

In the mid-1980s, Jaron Lanier pioneered the way when he began selling innovative VR software, goggles, and gloves. Unfortunately, Lanier was still several decades ahead of his time, and his company went bankrupt. Palmer Luckey, the founder of the VR start-up Oculus, picked up where Lanier left off. In 2012 Luckey's crowdfunding campaign on Kickstarter for a VR headset raised millions of dollars and captured the public's imagination. Soon, everyone was talking about the future of virtual reality. After landing venture funding but before the product ever shipped, Luckey was lucky enough to sell his start-up to Facebook for billions of dollars—further proof that timing is everything.

On the heels of Oculus came a wave of third-party devices that promised to make the experience even more immersive. The first thing most people want to do in a VR environment is walk around, but that's hard to do without bumping into furniture, hitting a wall, or tripping over some object. A small industry has sprung up around creating VR treadmills that promise the freedom of movement you would expect in a virtual environment. These treadmills are designed to enable users to run, duck, twist, sit, and jump inside a virtual world.

A different VR gadget that's more portable is the haptic glove. These provide sensory feedback to your hands, enabling users to feel the shape, texture, and motion of virtual objects. There are also devices that don't require any gloves but can track the movements of your hands. These use infrared cameras and LEDs to translate your hand and finger gestures into the virtual world.

If you really want to go all out, there are haptic suits. These facilitate a full-body VR experience. They look like something out of *Ready Player One*. Some suits even work for motion capture and include sensors for recording users' vitals and emotional stress levels. To immerse yourself further, you can don a multisensory mask that simulates hundreds of smells, ranging from gunpowder and burning rubber to freshly baked bread.

The problem with all these accoutrements, and the reason they are still an ultra-niche market, is that they are expensive and awkward. High-end haptic gloves alone can set you back a couple of thousand dollars. They are also a pain to wear. Imagine the amount of work required to put on all this gear. And then there's the issue of compatibility. Not all VR games and apps work with these devices. So, you just may find there's no aroma pack for your favorite game.

If a new product is harder to use than existing solutions, most people won't bother to switch. No one wants the hassle. Right now, smartphones are tough to beat. If you want to play a game, order lunch, or check your messages, it's just a tap away. VR, on the other hand, requires you to wear a cumbersome mask and spend time navigating a tricky user interface before you can even begin to enjoy the benefits.

The sheer amount of work and time necessary to perform basic tasks is what has slowed mass adoption. It's also strange and disconcerting to block out the real world in a public space. People don't like to use VR devices when outside or in cafés. That means most VR activity takes place at home or in a specialized environment. This severely limits its usefulness when compared to a smartphone.

Keep in mind that all these barriers won't matter if virtual reality can deliver upon Heilig's original promise: as soon as people can actually step "inside" a movie, virtual reality will be irresistible. But for this to happen, virtual reality needs to deliver novel value to consumers that other forms of entertainment can't match. Imagine being able to live out your wildest fantasies inside worlds that are so rich and compelling they feel lifelike. It will be an entirely new type of mass connectivity, where people will invest huge amounts of time, money, and energy into creating their virtual identities. The relationships and interactions they have in these virtual worlds will become as meaningful and essential to them as anything in their "actual" lives—maybe more so.

In fact, there may come a time when people no longer distinguish between what's real and what's virtual. Even the term "virtual reality" may disappear. That's how we'll know when virtual reality has finally arrived.

ARTIFICIAL SENSORY PERCEPTION

Our brains are black boxes. They don't know anything about the outside world except what our senses tell them. Our senses translate everything we see, hear, smell, touch, and taste into electrical impulses. These are fed into the brain, and we learn over time to interpret them. What most people don't realize is how malleable our brains actually are. If you change the incoming signals, you can change what we perceive as reality.

In the future, instead of bothering with all the VR gear, we will probably use some form of advanced interface to send electrical signals that impersonate sights, sounds, smells, tactile sensations, and tastes directly to the parts of our brains where those types of input are processed, and we would interpret them as real.

This sounds like a fantasy, but you may be surprised to learn that we aren't that far off. Deaf people can receive cochlear implants that allow them to hear. It's a surgically implanted neuroprosthetic that bypasses the normal acoustic hearing process, replacing it with electric signals that directly stimulate the auditory nerve. More than three hundred thousand hearing-impaired people around the world already have cochlear implants.

Not only can we do this with hearing, but we can do it with sight. A start-up called Second Sight has developed Orion, a brain implant that converts images from a camera mounted on a pair of sunglasses into electric impulses that stimulate the brain's vision receptors. The artificial vision is far from perfect, but it does allow the blind to see again.

"I still can't put it into words. I mean from being able to see absolutely nothing—it's pitch black, to all of a sudden seeing little flickers of light move around," says Jason Esterhuizen, who lost his vision in a car accident. "This is really the first time that we've had a completely implantable device that people have been able to go home with," explains Nader Pouratian, a surgeon at University of California, Los Angeles.[14]

What about the sense of smell? Scott Moorehead suffered a traumatic brain injury when he fell in the driveway while teaching his six-year-old son how to skateboard. He recovered from the internal bleeding and concussion but never regained his sense of smell. The connection between the olfactory nerves in his nose and his brain had been severed.

"Until you can't smell at all you have no idea how emotional the experience can be," said Moorehead, who fell into a deep depression. "You start to think about these really awful things, like someday my daughter is going to get married and I'm going to walk her down the aisle and I'm going to give her a big hug, and I'm going to have no idea what she smelled like."[15]

Despite his depression, Moorehead isn't one to give up easily. He became a successful businessman running the largest Verizon retailer in the nation, and he was determined to improve his condition. Through a friend, he found out about a team at Virginia Commonwealth University working on converting chemical scents into useful electrical signals. Moorehead offered to become not only their test subject but also their business partner. He wound up supplying the initial capital to kick off the commercialization efforts. Together, they've launched a start-up called Lawnboy Ventures, based on a popular Phish song. Moorehead hopes this will not only bring back his own sense of smell but help millions of others in the same position.

Would you like to taste something in a virtual world? At the National Institutes of Health, scientists found that taste doesn't require a tongue at all. Stimulating the taste cortex of mice was enough to trick them into thinking they'd tasted sweet or bitter substances. To prove this, they implanted a fiber-optic cable in the brains of mice that turns neurons on with laser light. By switching on the bitter sensing part of the insular cortex, they were able to make mice pucker up, as if they'd eaten something bitter. In a second experiment, the researchers fed the mice a bitter flavor but then made it more palatable by switching on the sweet sensors in the brain.

"What we discovered . . . is that there are regions of the brain, regions of the cortex, where particular fields of neurons represent these different tastes again. So, there's a sweet field, a bitter field, a salty field, etc." explains Nick Ryba, a sensory neuroscientist.[16]

Instead of implanting a fiber-optic cable in the brain, which most people would be reluctant to do, Homei Miyashita, a researcher at Meiji University in Japan, has taken a different approach. He's created an artificial flavor generator small enough to stick in your mouth. This device uses five different gels, corresponding to the five tastes the human tongue can distinguish: salty, acidic, bitter, sweet, and umami. He calls it the Norimaki Synthesizer, and he says the popsicle-sized gadget can reproduce any taste.

"Like an optical display that uses lights of three basic colors to produce arbitrary colors, this display can synthesize and distribute arbitrary tastes together with the data acquired by taste sensors," says Miyashita. The device fooled test subjects into believing they were tasting "the flavor of everything from gummy candy to sushi without having to place a single item of food in their mouths."[17]

Miyashita believes this technology could add "a whole new medium to multimedia experiences." In other words, get ready to suck on Norimaki Synthesizer if you want to taste what you see in a virtual world. Either that or get a brain implant?

Most healthy people today would never consider implanting a chip inside their brains, but in the future, this might not be the case. At some point, we'll have incredibly tiny chips and robots that can be easily implanted into the brain or injected into the bloodstream. These nano-sized devices would be resistant to corrosion from biofluids, so they could remain functional for years and allow us to experience lifelike virtual experiences by stimulating regions of our brains directly.

Inventor, author, and scientist Ray Kurzweil believes these robots would "go into the brain and provide virtual and augmented reality from within the nervous system, rather than from devices attached to the outside of our bodies." Kurzweil has been pushing the limits of technology his entire life. As a boy, he fashioned telephone relays into a calculator that could compute square roots. At fourteen he came up with software that analyzed statistical deviance, which IBM distributed with the IBM 1620. After graduating MIT, he went on to invent the first charge-coupled flatbed scanner, optical character recognition software, and full text-to-speech synthesizer, among other things.

Kurzweil's team at Google is now working on rough simulations of the neocortex. No one has a perfect understanding of the neocortex yet, but engineers are already able to do interesting applications with language. By the 2030s Kurzweil believes they'll have very good simulations of the neocortex.

"The most important application of the medical nanobots is that we will connect the top layers of our neocortex to synthetic neocortex in the cloud," explains Kurzweil. "Just as your phone makes itself a million times smarter by accessing the cloud, we will do that directly from our brain. It's something that we already do through our smartphones, even though they're not inside our bodies and brains, which I think is an arbitrary distinction. We use our fingers and our eyes and ears, but they are nonetheless brain extenders. In the future, we'll be able to do that directly from our brains."[18]

It will probably be the combination of nanotechnologies and advanced brain-computer interfaces that enable lifelike virtual experiences. Creating ultrarealistic simulations, like those depicted in *The Matrix*, will happen, and when it does, large numbers of people will take the leap. They may not even need to submerge themselves in sensory deprivation tanks to engage in a full-sensory virtual experience. Just sitting on a chair or lying down in bed may be enough. The brain implants would simply block any signals coming from the rest of the body, enabling the person to entirely transition to a virtual state of being.

When this comes about, the draw will be irresistible but also frightening on an existential level. People may not want to return to their real lives. We've all heard of those unfortunate souls who have died because they couldn't stop playing video games. A simulated universe might become so addictive that many people would opt to be placed on life support just so that they don't have to disengage. At this point, what happens to the human race? Will we allow our machines to carry on without us, or will we program our robots to continue to produce more humans using advanced in vitro fertilization techniques and artificial uteruses?

In the end, what's the point in continuing the human race if all we want to do is escape into a simulation? That's a question for philosophers, but honestly, I don't think we'll go that far. There will always be people who view disappearing into a virtual world, no matter how realistic, as a hollow, meaningless experience. They would rather do something with their lives that they feel is valuable and has purpose. These people won't opt for the enchantments of a virtual life. There will also be those, who for religious or personal reasons, refuse to enter any simulation. They will remain on the outside, living much as we do now. Maybe they will miss out on the wonders of a billion simulated worlds, but they will also be the ones to carry on our journey in this world, even if it is only another simulation.

IS THE WORLD A SIMULATION?

At the Recode's 2016 Code Conference, Elon Musk said, "Forty years ago, we had Pong—two rectangles and a dot. That's where we were. Now, forty years later we have photorealistic, 3D simulations with millions of people playing simultaneously, and it's getting better every year. And soon we'll have virtual reality. We'll have augmented reality. If you assume any rate of improvement at all, then the games will become indistinguishable from reality, just indistinguishable."[19]

Musk went on to say that either we will make simulations that we can't tell apart from the real world "or civilization will cease to exist." He even believes we may already be living in a virtual simulation, saying, "There's a billion to one chance we're living in base reality."[20]

Musk isn't alone. Scott Adams, the creator of the Dilbert comic strips, also thinks we may be living in a simulation. He wrote on his blog, "When I was a kid, I dreamed of one day growing up to be a world-famous cartoonist. When your actual life conforms to your childhood fantasy, it makes you question the basic nature of reality. Did I really beat million-to-one odds, or is something else going on?"[21]

Adams points to the physics behind his beliefs. First, if we are actually living in a simulation, we should expect that we cannot travel past the boundaries of the simulation. This is, in fact, true. No one can travel beyond the edge of the universe without exceeding the speed of light. Second, we would not be able to observe the basic building blocks of reality. This, also, is true. When we drill down to the quantum level, everything is probability. We can't be certain of anything in the quantum world, not even the position and momentum of a photon at any single point in time. Adams points out that it's similar to many computer games, where the boundaries to the universe are impossible to reach, and an algorithm dynamically generates the world as users play.[22]

Nick Bostrom, a Swedish philosopher at the University of Oxford, wrote a paper called "The Simulation Argument" in which he contends that if humanity does not go extinct before it reaches an advanced stage capable of superintelligence, and if this civilization is inclined to develop simulations that re-create the lives of its ancestors, then we are almost certainly living in one of those simulations. Why? Because the cost of making a lifelike simulation will be so low in an advanced civilization that there will be billions of these simulations, and the odds are that you exist in one of the billions of simulated realities, as opposed to the one true reality.

Does all this mean we are living in a simulation? I don't think so and neither does Bostrom. It merely points to the possibility that it may be true. My personal opinion is that if we are living in a simulation, someone could have done a better job. I'd certainly have designed it differently. But since we don't get to choose our simulations, I guess we're stuck with whatever random one we were handed. The only consolation is that there are probably other versions of us in much better and worse realities, which jives nicely with the multiverse of string theory. But we won't get into that now.

Suffice to say that we can never know for certain, so you are welcome to believe whatever you want. What we do know is that creating near lifelike simulations will be possible at some point in the future, and when this does happen, people will flock to them. Why wouldn't any of us indulge ourselves, especially if we could live out our wildest fantasies? It would be a blast for most people to transform into anyone or anything they could imagine and go on realistic, hair-raising adventures, without any risks of winding up permanently traumatized, disfigured, or dead.

When we get to this point, I suspect many people will vastly prefer the simulation over their actual lives and abandon the real world altogether. But would this be such a bad thing? It really depends on how much you value reality or what you decide reality is. If you believe reality is already a simulation, then it wouldn't be a problem. You'd just be abandoning one simulation for another.

HYPERLAYERS: AUGMENTED REALITY

Lifelike simulations aside, augmented reality is poised to have an even bigger near-term impact on our lives than virtual reality, and that's because it takes the digital world and combines it with the physical world. Unlike virtual reality, where the user enters a computer-simulated world, augmented reality (AR) is an interactive experience of the real-world where the physical environment is visually, audibly, and digitally enhanced.

Like with virtual reality, AR has had some false starts, the most publicized of which was Google Glass. There was a lot of hype around this little device that could layer digital information over the real world, but the experience left a lot to be desired. The tiny screen, awkward interface, and limited utility made the device more of a gimmick than a breakthrough. The camera was what sealed its fate. Many people, most of whom never tried the device, detested being captured on video without their permission and dubbed the early adopters "glassholes."

The mobile game *Pokémon GO*, on the other hand, was a phenomenal success. Downloaded more than half a billion times in the first year alone, it demonstrated the future promise of augmented reality. The game is simple to play: just hold up your phone's screen and watch as animated monsters appear on top of whatever your camera is seeing. The goal is to capture these monsters and train them to fight other players' monsters. But is *Pokémon GO* truly AR? Most experts consider it pseudo-AR. Holding up your phone is not an immersive experience. It's more like the halfway point between a full AR experience and a phone-based application.

Many other apps have tried using the same technique and failed. Most people simply don't want to hold up their phones to use an app or play a game. It's tiresome and the experience is subpar at best. For AR to truly go mainstream, it has to come in an entirely different form factor. Magic Leap promised this when they launched their AR headset, but it was too cumbersome for most people. Apple, Google, and others are working on lighter-weight headsets and glasses that can deliver the type of immersive experience people expect out of AR, but no one has cracked the code yet.

Despite the technical hurdles that remain, AR is progressing at a remarkable pace. There are an estimated hundred million people using AR-enabled shopping technologies right now, and by 2025 the global market for VR and AR in retail is expected to reach $1.6 billion. AR applications for indoor navigation are another area poised for rapid growth. Today, most people use Google Maps and similar applications to find directions outdoors, but when they are inside a shopping mall or office complex, getting to the right location quickly can be tricky. This will change as people increasingly use AR applications on their phones to guide them.

The automakers are also bringing AR navigation into cars with heads-up displays. We'll no longer have to look down at a screen to get directions. Instead, the image will be projected onto the front windshield. We will not only see directions superimposed on our field of view but also hazard warnings, traffic alerts, and dashboard information, like driving speed, fuel gauge, and weather conditions.

The biggest area for AR growth in the near term is industrial applications. Mixed-reality promoter VRX conducted a survey for its XR Industry Insight Report 2019–2020 that indicates 65 percent of AR companies are now focused on industrial applications.[23] Forrester estimates that fourteen million American workers are expected to use smart glasses regularly on their jobs by 2025. This will include everyone from factory and warehouse workers to designers and dentists. AR is also being used by technicians and maintenance staff to quickly fix problems and make repairs, without having to consult an expert or read lengthy manuals. Already, AR-based training in enterprises is big business. ABI Research estimates the market at $6 billion in 2022.[24]

The health-care industry is probably the fastest-growing segment. AR adoption is forecast to increase by 38 percent annually until 2025.

Augmented reality apps are being used by surgeons during operations to alert them of risks and hazards, while providing detailed information on patients and procedures. Other applications help nurses find patients' veins and avoid mistakes. There's even an AR app that is designed to show anyone using a defibrillator exactly what they need to do in an emergency.

Given all this progress, what will it take for an AR device to become as commonly used as our phones? First, we will need to develop hardware that doesn't get in the way. The device must be something people don't mind wearing on their faces for extended periods. Glasses have potential, but most of us would rather not wear them. Another option is contact lenses. A number of companies, including Mojo Vision, Google, and Samsung, are working to develop AR micro-lenses that are thin and comfortable enough to place on your eye, but this is no easy feat.

An even bigger challenge than the hardware is the user experience design. For AR to take off with consumers, it has to be incredibly simple and intuitive to use. We need to figure out a way for people to effortlessly navigate and interact with virtual content superimposed on the real world. Only after we overcome the usability issues will the majority of people embrace AR devices.

Once this happens, augmented reality will fundamentally transform how we live and work. It will be the equivalent of turning the entire world into an extraordinarily powerful web browser. Objects will suddenly possess entirely new dimensions and functionality. Let's say you pick up a bottle of prescription pills. All the information can be available right before your eyes in high-definition video, infographics, and 3D models. The system will identify the medication, warn you of any negative side effects or interactions, and instruct you on the dosage.

Imagine a hyperconnected world that is layered with all the information you need just waiting for you to access it without fumbling for a phone or laptop. Care to shop while traveling in your self-driving car? Just bring up a virtual store and begin browsing. Thinking of booking a hotel in Hawaii? With the wave of your hand, you can walk through the resort in 3D. Need sales materials to show someone at a conference? Up pops a virtual representation of your product ready to demo.

Once this is possible, what we think of as mass connectivity will entirely change. Instead of staring at our phones for hours every day, we'll be looking up at a world around us, while engaging with seemingly endless layers of information and interactive digital content woven into the fabric of our everyday lives.

MIXED REALITY: LIVING A MULTIMODAL EXISTENCE

As mass connectivity evolves, we'll probably wind up living a multimodal existence, where we can transition seamlessly between the physical world and augmented and virtual realities. Brain signal manipulation would make virtual objects not only look real but feel real, activating all five of our senses. In the same way that a brain-computer interface could enable lifelike VR simulations, it could merge the virtual with the physical into a mixed reality.

In a complete multimodal existence, the dividing line between what is a digital and material object will diminish. If I can pick up a virtual laptop, feel its form, texture, and even heft, it will become not only real but really useful. I'd prefer a virtual laptop that I didn't have to lug around with me over a physical one, especially if I could get realistic tactile feedback from the keyboard. I'd even pay more because of its convenience.

Imagine walking through an augmented city park, where you can reach over and not only touch the virtual flowers but smell them. We may even be able to buy virtual snacks at the park, like a bucket of caramel-covered popcorn, and get a sense of the texture and flavor without the calories. During our stroll, we may also encounter objects that may have never existed in the real world, such as virtual buildings and enhanced landscapes. There may even be a virtual park ranger with a charming AI personality that is available to assist us.

In a multimodal existence, we are likely to spend a great deal of our time customizing our own virtual environments. Instead of purchasing actual pets, paintings, and potted plants for our homes, we could acquire virtual ones. Why? Because they would not only look realistic but have additional benefits. We wouldn't need to worry about caring for our pets, our paintings could easily be updated, and our virtual plants could grow without any natural sunlight or water.

It won't only be homes that get decorated. Virtual facelifts and body modifications may become commonplace. As with today's beauty apps, teenage girls may get creative, making their eyes larger and lips fuller, while sporting devil horns on their heads and stars on their cheeks. But probably it won't stop there. They may also modify their bodies with catlike tails, glowing skin, and animated tattoos.

People will probably spend inordinate amounts of time designing their worlds because they will embody who they are and how they want to be perceived by others. Some folks may even turn world building into a full-time job. Just like in today's computer games, there will likely be rare collectibles and branded goods. Entire new industries will develop around creating and designing virtual objects, attire, and environments, with famous fashion designers, architects, performers, and world creators competing for consumer dollars.

When multimodal existence becomes a part of daily life, users who don't want to experience a mixed reality will probably be able to turn it off, but in doing so, they will lose access to much of the functionality. It would be worse than leaving your smartphone at home for the day.

For those who opt in, the world will become a much more fanciful place. There will be AI-controlled avatars working in shops, taking our orders, and entertaining us. Different parts of the city may have different augmented overlays. In the historic zones and museums, everyone may appear in period attire, while at a sporting event, fans could paint the sky with their team colors. And at rock concerts, people may choose to participate in mass shared experiences, like virtual light shows.

The rules we play by may also depend on where we are. In shopping malls and commercial districts we may have to put up with mixed-reality ads. At school, students might be forbidden from displaying virtual body augmentations or writing virtual graffiti on the walls. When in the office, companies may manipulate our reality to increase collaboration, communication, and productivity. And when we walk into a courthouse or police station, the system may override our personal settings, blocking any sort of transmissions.

This may also be true when we travel to foreign countries. I can see a future where Asian and Middle Eastern cultures require greater conformity than their Western counterparts. Self-expression might be censored or limited. And the dominant platforms that control what we experience may change, resulting in vastly different environments. We will inevitably have to adapt to the local rules and social norms. This means there probably won't be a universal mixed-reality experience but countless separate realities depending on where we are, what we're doing, and which platforms we're using.

THE SEVENTH WAVE OF MASS CONNECTIVITY

Throughout our history, mass connectivity has come in waves, each one permanently transforming human civilization. It began with spoken language, followed by written language, the printing press, telecommunications, the television, and the internet. We are now on the cusp of the seventh wave of mass connectivity.

When it arrives, it will make it far easier to tap into information and use it in our daily lives. Implanting chips in our brains and entering an augmented reality may seem strange, even frightening, but we have been augmenting our bodies since our prehistoric ancestors first learned to use stone hammers, wooden spears, and other tools. Building machines that extend our capabilities is how we have constructed our advanced civilization, and we won't turn back at the opportunity to connect our minds to the most powerful tool in human history: the internet.

What will make true multimodal mass connectivity possible is an exponential increase in performance. We have to remember what happens when new technologies are deployed at scale. The early supercomputers filled entire buildings, and they were orders of magnitude less powerful than what we can wear on our wrist today. Sequencing the first human genome cost around $1 billion and took thirteen years to complete. Today, it costs less than $1,500 and takes hours to complete. It won't be long before an advanced brain-computer interface and augmentation system becomes less conspicuous than a pair of Bluetooth earbuds and far more capable than anything that exists today.

Once our brains are connected and capable of full sensory augmentation, humans will enter a new phase of our existence, where the physical and virtual merge into a single reality. This will not only change how we interact with the world but enable us to operate at a much higher level. Imagine exchanging not just messages but thoughts and emotions with other human beings.

We will no longer be trapped inside our own bodies. For the first time in human history, we will be able to know what it's like to tap someone else's consciousness and share our lives in a way we have never experienced before. The gap between our mental, physical, and digital states and that of other human beings will disappear. At this point, mass connectivity will have achieved what humans have dreamed of for millennia. We will have transcended ourselves.

No one knows if this will come about in twenty or two hundred years, but the one thing we do know is that it's coming, so hang on: It's going to be a wild ride!

FORCE 2

BIO CONVERGENCE

Bio Convergence, the force driving humans to merge biology with technology, will enable us to decode the building blocks of life, create entirely new species of plants and animals, conquer disease, and heighten human abilities.

We are entering an age where biology and technology are converging, enabling everything from lab-grown beef to bionic eyes and ears. With the sequencing of the human genome and the advent of gene-editing tools, humanity is poised to reinvent the natural world. By manipulating the source code of life, we can now create pest-resistant crops, faster-growing fish, and livestock that can survive in extreme conditions. Genetic engineers have even begun to eliminate previously incurable diseases and develop life-forms that never before existed.

The natural limitations of our bodies are no longer immutable barriers. Biohackers are experimenting with embedding microchips in their arms, legs, and chests to monitor their health, enhance their senses, and access information. Mind hackers are popping smart drugs to try and boost brain power. Start-ups are offering gene therapies and infusions of young blood in an attempt to extend life spans, while cryogenic facilities are freezing bodies and brains so that people could potentially be brought back to life at some future date.

This brings us to the moral questions surrounding these technologies. Is it ethical to genetically alter a fetus to prevent deadly diseases? What about using gene-editing tech to enable designer babies with genius IQs and movie-star good looks? Should we create transgenic pigs with human DNA to act as organ donors? Where do we draw the line when it comes to playing God with nature?

It's up to all of us to decide what role technology should play and which path to take as we pioneer this uncharted territory. We'll discuss this and much more as we enter the wild new world of bio convergence.

WETWARE WARRIORS: BIOHACKERS ON THE LOOSE

What's different about biohackers is their mindset. These self-described insurgents aren't waiting for governments or universities to give them the greenlight to experiment. They believe there's no need to hold off for double-blind, randomized, placebo-controlled trials or any sort of institutional approvals. Hackers tend to see themselves as trailblazing rebels and want to push the limits of biology entirely on their own terms. They are willing to work out of garages, basements, and makeshift labs, run their experiments on a shoestring budget, and are open to trying just about anything—even if it breaks a law or two.

Many biohackers call themselves transhumanists, and they believe humans no longer need to accept the biological limitations imposed by nature. Instead, humans should engineer past these barriers using a range of new technologies, like gene-editing kits and electronic implants, as we augment and evolve our species. They often draw their inspiration directly from sci-fi books and movies and see themselves at the vanguard of a brave new world. These supergeeks are determined to enhance their senses with electronics, boost their intelligence with drugs, alter their body parts with implants, and recode their DNA with gene-editing kits. In short, they want to become the first transhumans in the coming posthuman world.

The term *biohacking* first appeared as early as 1988 and was used to describe amateur engineers running experiments with organic matter in DIY labs. Since then, the term has broadened as biohackers have gone on to organize their own movement called DIY bio, which stands for do-it-yourself biology. There are now DIY bio groups in major cities across North America, Europe, and Asia. These biohacking clubs include Berkeley BioLabs, Genspace in New York, Open Wetlab in Amsterdam, and La Paillasse in Paris. Many of them offer equipment and classes, and some even provide funding.

For the most part, it's hobbyists, amateur scientists, and entrepreneurs doing basic experiments similar to what you'd see in an advanced biology class. This includes projects like developing glow-in-the-dark yogurt, creating genetically altered frogs, and finding ways to turn vitamin-rich algae into tasty drinks.

"The era of garage biology is upon us," says Rob Carlson, managing director at Bioeconomy Capital and a pioneer in the space. "The predominant thought about biology used to be that it was expensive and hard. And it's still hard. It's just not so expensive."[1]

Given that a basic gene-editing kit can be bought online for as little as $100, the field is wide open to anyone with the knowledge and a lot of time to fool around. It no longer costs a small fortune to acquire clean proteins and other basic materials. Even the cost of lab equipment is coming down, as hackers build their own low-cost instruments, like molecular photocopying machines and labs on a chip. Today, with just a few thousand dollars, garage biohackers can get started.

BioCurious, located in the heart of Silicon Valley, caters to these aspiring innovators. Its website proudly states that it is the world's largest community lab space for biology. Its projects include developing an open-source bioprinter, cuttlefish ribonucleic acid sequencing, ribosomal sequencing of bacteria and yeast, and a multiweek hackathon to build an inverted, optical, fluorescing microscope. Sound fun? Apparently, a lot of people think so. There are tens of thousands of biohackers out there, and the number is growing quickly.

Biohacking is not just limited to science geeks and entrepreneurs. There are a number of artists and designers experimenting with biohacking. As early as 2000, Eduardo Kac, a self-described transgenic artist, collaborated with the Institut national de la recherche agronomique to produce a green-fluorescent bunny. He accomplished this by implanting a protein gene from a jellyfish into a rabbit, which made it glow green under ultraviolet light. Whether you feel this is art or not, it's definitely biohacking.

For those of us who don't have the time or inclination to set up our own lab but would still like to invent a species of plant or animal that never existed before, we can hire Cambrian Genomics. Operating out of a makeshift laboratory in San Francisco, this synthetic biology start-up lets customers tinker with the genetic codes of plants and animals and even design their organisms on a computer.

"Anyone in the world that has a few dollars can make a creature, and that changes the game," says CEO Austen Heinz. "And that creates a whole new world."[2]

The company's clients include Petomics, which is making a probiotic for cats and dogs that makes their feces smell like bananas, and Sweet-Peach, which is developing personalized microorganisms to promote vaginal health.

GRINDERS AND CYBORGS

Where it gets dicey is when biohackers start experimenting on themselves with the goal of upgrading their bodies and brains with the latest technologies. These highly adventurous individuals would rather not wait for science to catch up to their favorite comic books, so they take it upon themselves to begin the process. This includes trying to extend one's physical and mental abilities beyond normal human limitations by inserting mechanical devices and electronics into their flesh, experimenting with untested drugs, conducting novel surgical procedures on their anatomy, and even going as far as to use gene-editing techniques to give them new powers.

Many biohackers are hipsters looking to join an edgy subculture, while others are motivated by the desire to be better than those around them. A few are driven by medical problems that can't be fixed using generally accepted methods. And let's not forget those eager inventors, who don't want to wait for FDA approval and don't mind becoming their own guinea pigs.

Whatever their motivations, self-experimentation is as old as science. In the early nineteenth century, Stubbins Ffirth, a medical student, determined to prove yellow fever was not contagious, smeared himself with tainted vomit, blood, saliva, perspiration, and urine. Ffirth even went so far as to ingest the black vomit straight from a patient's mouth. He didn't get yellow fever but did receive his medical doctorate.

Another intrepid soul was German surgeon August Bier, who in 1898 discovered spinal anesthesia. He experimented first on himself and then on his assistant. To prove the spinal anesthesia worked beyond a doubt, Bier stabbed, hammered, and burned his assistant, and then proceeded to pull out his pubic hairs and squeeze his testicles. It shouldn't come as a great surprise that his assistant eventually left him when they couldn't agree on who was ultimately responsible for the invention of spinal anesthetic.

I could go on with many more examples from the past but suffice to say that self-experimenting biohackers are following in a long tradition and see themselves as daring pioneers. There is no better example of this than Tim Cannon, a software developer from Pittsburgh and cofounder of Grindhouse Wetware. His team likes to say computers are hardware, apps are software, and humans are wetware.

"Ever since I was a kid, I've been telling people that I want to be a robot," says Cannon. "These days, that doesn't seem so impossible anymore."[3]

Cannon doesn't just talk the talk. Back in 2013 he inserted a hefty battery-powered device into his arm. Because no board-certified surgeon would perform the operation, he had his DIY team, which included a piercing and tattoo specialist, slice open his skin and insert the device, then sew it back up. The photos look gruesome, but Cannon was satisfied with the results. He called it the Circadia 1.0, and it recorded bio data, like body temperature, which was then transmitted directly to his smartphone.

Cannon and his buddies at Grindhouse Wetware went on to develop a glowing disc they implanted into the back of their hands. This device, "much like piercings and cosmetic surgeries, is purely for aesthetic purposes," explains their press release. "It is a simple device that will prove the possibility of implanting technology in the body and will pave the way for more advanced and functional augmentations."[4]

Cannon and his cohorts aren't alone. There are hundreds of people augmenting their bodies by surgically implanting electronics and various materials under their skin. They like to call themselves grinders and gather in online forums like biohack.me, where hundreds of wannabe cyborgs exchange tips on surgical techniques, how to get illegal anesthetics, and the best bioresistant coatings.

Another leading biohacker is Kevin Warwick, a professor at Coventry University. Warwick has been implanting chips in himself for decades, earning him the nickname Captain Cyborg. In 2002 he connected an electronic implant to the nerve fibers of his left arm, then plugged this into a computer which monitored the nerve signals from his brain to his arms, receiving and transmitting them like radio waves.

In one experiment, Warwick flew to New York and hooked up his cyborg arm to a computer, which used the internet to connect it to a robotic hand back in England. When he moved his hand, the robotic hand mimicked his gestures. The robotic hand then sent signals back to Warwick so that he could feel how much force it was applying to any object. The result was that his body and brain didn't have to be in the same place. He could extend his consciousness around the globe.

Warwick went on to convince his wife to implant a microchip in her wrist without any anesthetic. This didn't end their marriage but instead brought them closer together, both physically and emotionally. By connecting her to the internet, Warwick linked his nervous system with his wife's. Whenever his wife closed her hand, his brain received an electric pulse. They began to communicate with each other electrically through their nervous systems. The two of them started blipping at each other all day long. This created a new type of intimacy between them.

"When you think of what's involved in sex, sex is quite intimate, but it's not this intimate. This really is! You're getting the insides of your body hooked up together," says Warwick.[5]

If you happen to be one of Captain Cyborg's students, be prepared for some unusual class projects, like turning one's fingers into antennae. One of his students wanted to be able to feel in his fingertips how far away an object was. By attaching an ultrasonic sensor to his head and running wires down around his fingers, he succeeded in transmitting distances as electrical impulses that made surgically implanted magnets in his fingers vibrate.

This gave him a sense of distance without seeing, which may someday be useful for blind people.

Sensory augmentation is at the heart of many of the biohacking upgrades. Liviu Babitz, who also wants to create new human senses, has inserted an electronic device into his chest that vibrates every time he faces north. He calls this biocompass North Sense and sees it as the first step in a complete bionavigation system. Babitz envisions a world where people can navigate like birds. Without even looking around, they will know exactly where they are at all times. It sounds kooky, but in a future where augmented reality becomes the norm and our vision is clouded by layers of digital content, this type of extrasensory navigation may prove useful.

Richard Lee, a salesman from Utah and father of two, has gone an entirely different direction. His goal is to enhance people's sex lives by upgrading their private parts. His first attempt at this is the Lovetron9000. When implanted under a man's pubic bone, it makes the penis vibrate.

"I'm not sure what made me want to develop the Lovetron," Lee admits. "After my divorce, I abandoned the project altogether, figuring I'd be celibate forever. But then I met a woman with a background in sex [psychology] who urged me to finish it. She convinced me that it would be empowering in many relationships."[6]

Lee's experimentation goes beyond the orgasmic. He has a near-field communication (NFC) chip in his hand for remotely controlling devices, a biotherm chip in his forearm for monitoring temperature, two magnets in two different fingertips to detect magnetic fields, and magnetic speaker implants in the cartilage of both ears. Because he's losing his sight in his right eye, he plans to hook the speaker implants up to an ultrasonic rangefinder so he can echolocate like a bat.

One experiment that didn't go so well for Lee was when he implanted polymer-foam tubes into his shins that he claimed could withstand the full force of a baseball bat swing. He called this his non-Newtonian armor. Unfortunately, his stitches burst open, forcing him to remove the tubes. But this hasn't stopped him from experimenting on himself.

"Accepting the stupid default given to us by nature is an unnecessary act of submission or compliance," asserts Lee.[7]

If it seems bizarre and even ghastly that amateurs are willing to slice open their shins, thighs, ears, fingers, arms, and even groins to become cyborgs, entire countries are now hopping on the biohacking bandwagon. Sweden started a trial using NFC implants for public transportation. Around 1,500 test subjects had tiny NFC chips embedded under their skin, enabling them to check in at train stations simply by swiping their hand.

In the future, having body implants may not only become practical but unavoidable. People may get body implants at birth and regular updates throughout their lives, just like we upgrade our phones and smartwatches. The devices will inevitably shrink in size so that painful surgeries become a thing of the past. Soon enough, getting electronics implanted might be easier and less painful than a tiny tattoo. Eventually, devices may become so small that implanting one will feel like nothing more than a pinprick. At this point, biohacking may not seem so extreme. We all may be grinders someday.

RESHAPING THE HUMAN ANATOMY

There's another form of biohacking that isn't about expanding a human's mental or physical capabilities. Instead, it's all about aesthetics. The biohackers are intent on transforming how they look, creating bizarre and often disturbing alterations to their faces and bodies. These hackers often wind up looking like they just walked off the set of *Star Trek*. They can have horns protruding from their heads, bumps along their arms, even hearts and stars rising up from under their skin on their chests and backs.

Using the latest tech, skilled body modification artists are in high demand. They are masters at sculpting, shaping, and illuminating the body in ways that were previously impossible. The first implants were stainless steel, but today they are mostly molded silicone. A body artist typically uses a scalpel to make a shallow incision in the skin and then switches to a dermal elevator to open up a cavity just big enough to slide in the implant. After this, the incision is sewn shut. Most people don't receive any anesthetic because it's hard to come by legally. So they just bear the pain.

Part of the allure for many body hackers is the shock value. There's a punk rock ethos to it. It used to be that getting a Mohawk and walking around in a leather jacket was enough to build your rebel cred. Then came tattoos and radical piercings of the nose, lips, face, and other body parts. Today these are so commonplace that the truly defiant and subversive are seeking more radical transformations. It's all part of making a statement.

"When you have technology and biomedical research and a pissed-off angry population that loves tattoos, this is bound to happen," says Cannon, when talking about why he started Grindhouse Wetware in Pittsburgh. "It's got the right amount of fuck you."[8]

But that's not the only reason people do this to their bodies. Members of the Church of Body Modification see themselves as practicing an ancient tradition with modern technology. They believe engaging in body modification rites and rituals strengthens the bond between mind, body, and soul, helping them to become spiritually complete individuals.

Others just like how it looks. They think it's cool, beautiful, and even sexy. A modification that may appear gruesome to some people can be attractive to others. There is a whole subculture around modification fetishes, with people going so far as to radically modify their genitals with implants, piercings, and surgical operations. I won't go into detail, but if you're curious, you can find pictures online. Just prepare to be shocked. If you can imagine it, someone has probably done it.

If all this sounds like a little too much, remember body modifications have a long history. Females with long necks are considered beautiful in many cultures. To elongate their necks, women in Southeast Asia started wearing thick metal coils around their necks as young girls, as far back as the eleventh century BCE. This same practice emerged in the Ndebele tribe in Africa, where women wear brass and copper neck rings. Each time they add more rings, it stretches the neck further.

In Mesoamerica it was common for people to drill holes in their front teeth and fill them with jade or iron pyrite. Some tribes took to filing their teeth into alternative shapes using stone tools. Modifying teeth with flat gold bands was practiced by Etruscan women in the seventh century BCE. And Viking warriors etched grooves in their teeth and filled them with red dye to look more fearsome.

The Chinese practiced foot binding as far back as the tenth century, when court dancer Yao Niang supposedly bound her feet to resemble the new moon. Emperor Li Yu appreciated this so much that it became a custom and sign of beauty. A three-inch foot came to be called a gold lotus, and was highly prized by potential husbands.

In this context, radical body modifications don't seem so unusual. They are just a natural extension of what humans have done for thousands of years. Even in modern societies, plastic surgery is considered normal. Surgeons routinely modify people's noses, cheeks, buttocks, and other body parts to make them look more attractive and socially acceptable. The biggest difference is that body modification artists are more experimental. They are willing to do things no plastic surgeon would ever attempt.

If someone wants Neanderthal-like implants under their eyebrows, a glowing bracelet inserted in their forearm, and a dozen marble-sized spheres embedded in their neck, why shouldn't they have it? After all, it's a personal choice and doesn't harm anyone, right? Not necessarily. Most body modification artists in America undergo no formal training. Many aren't experienced, and all of them lack government certification. It's the Wild West out there. Anyone can become a body modification artist just by hanging up a shingle.

Is this legal? Not really. They are operating in a gray area, where they do pretty much whatever they want. Some don't even think it's necessary to use professional instruments and will resort to using household implements, like butter knives, to separate the layers of skin. This is not only scary but irresponsible. In many procedures, there is serious risk of damage to the nerve and lymphatic systems, which can be a lifelong nightmare. And if the operation isn't done in a perfectly sterile environment, rates of infection skyrocket.

The problem is that even if you have the money and find a plastic surgeon sympathetic to the idea of implanting horns on your head, this professional won't do it. Why? Because no plastic surgeon would risk losing his or her license over such an operation. Like it or not, the American Medical Association (AMA) states you can't modify the body away from what society says is normal. The AMA is trying to reduce the risk of lawsuits and place some guidelines on what doctors can and cannot do. People often come to a plastic surgeon asking for something and then regret their choice weeks, or even years, later. These are tricky ethical and social issues, and the line has to be drawn somewhere. The result is that plastic surgeons dare not deviate from anything outside the norm, and by definition, radical body modifications are abnormal. So anyone who wants an extreme modification is forced to look elsewhere.

No matter what you think of this practice, just like tattooing is now mainstream, artistic and even bizarre body modifications could eventually become commonplace. Today's subculture is tomorrow's pop culture.

In decades to come, as technology progresses, human beings may modify themselves in even more extreme ways. Instead of simple implants, the future cyberpunks may opt to have an extra finger added or even an appendage. And they won't stop there. What about adding wings or even webbed feet? It may seem strange now, but what we perceive as beautiful, erotic, spiritual, rebellious, and even ordinary is constantly in flux.

If you live long enough, get prepared to meet people who may not even look human. We may find ourselves living in a *Star Wars* universe, where all the aliens happen to be modified versions of us.

SMART DRUGS, ENERGY BOOSTERS, AND SUPER SUPPLEMENTS

You don't have to slice open your body or inject yourself with an RFID chip to become a biohacker. There's an entire world of body transformation that involves simply popping pills. People are doing this at ever-increasing rates, with the goal of losing weight, gaining muscle, increasing endurance, improving energy, sleeping better, or becoming smarter.

Let's focus on one of the most interesting and fastest-growing areas: nootropics, which are drugs, supplements, and other substances that may improve cognitive function, including memory, creativity, and motivation. Have you ever dreamed of being the next Einstein or *Jeopardy* champion? There's a pill for that. The idea is that you can dramatically boost your brainpower simply by taking the right supplements.

Nootropics took off in hypercompetitive Silicon Valley, where geeks and entrepreneurs are always striving to maximize their productivity and boost their brainpower. Dawn Currin, a business analyst for an IT company, sums it up well: "We went as far as we could with caffeine. Everyone was getting tired of Red Bull. And we started drinking yerba mate. And someone was like, 'Hey, nootropics. Let's check what that's all about.'"[9]

"One may help bolster memory; another will help you focus. One pill helps improve vision, and another promises more energy," says Dave Asprey, a leading figure in the nootropics movement and CEO of Bulletproof Coffee. "They all have the same goal—to help you maximize your potential."[10]

Asprey, a techie and entrepreneur, once weighed more than 300 pounds and had trouble focusing at work. Now, due to his "brain-boosting" diet, he's a Silicon Valley superstar. After taking a fistful of pills every day and spending more than $1 million on smart drugs, he claims to have raised his IQ by twenty points. Given that the average IQ score is 100, a 20-point increase is substantial. Genius level IQ is considered to be 140 points and above. Keep in mind that IQ tests are not necessarily an accurate gauge of human intelligence and outcomes can vary from test to test. Also, there have been no clinical trials backing up Asprey's claims.

Despite skepticism as to the scientific validity of his methods, Asprey has become one of the best-known biohackers. This is because he's a tireless self-promoter. With broad shoulders and protruding biceps, he looks like a middle-aged triathlete. Whenever he gives a talk, which he does quite often, it's upbeat and inspiring—the type that makes people believe in themselves. He not only pushes his diet, brain boosters, and coffee but doles out advice on how to live.

Asprey has succeeded in combining traditional self-help with the latest technologies. He is armed with a cryotherapy chamber, an atmospheric cell trainer, a bed of infrared lights, a vibration platform, and an array of high-tech exercise machines, some of which claim to deliver two and a half hours of exercise in about twenty minutes.

"Biohacking is the art and science of changing the environment around you or inside you so that you have full control of your own biology," says Asprey.[11]

Like most people in the quantified-self movement, Asprey is obsessed with data. He uses implants, sensors, fitness tracking software, and even brainwave-measuring devices to capture as much data as possible. He is constantly monitoring his body, from heart rate and hormones to mental concentration levels, and then tweaking what he takes in. It's a process of continual experimentation.

At one point, Asprey wanted to test out Napoleon Hill's idea of sexual transmutation, where a person can take energy that the body wants to spend on sexual reproduction and harness it to change one's life. To do this, he used the Taoist formula, which says take your age in years and subtract seven, then divide by four. The result is the number of days a man should abstain from ejaculating in order to stay healthy and energetic.

At the time, Asprey wound up with the number eight, which meant that if he abstained from ejaculating (not necessarily sex) for eight days at a time, he would be healthier and happier. However, if he really wanted to go for it, the Taoist teachings recommended abstaining for thirty days. With his wife's permission and full cooperation, he began testing this theory. To do this, he graphed his daily happiness level with his frequency of sex and frequency of ejaculation. Except for the occasional mishap, the experiment went surprisingly well. His wife loved the sex because his stamina and desire shot up, and he became more energized than he ever was before. He now teaches this as part of his program.

"To this day, the Taoist equation is one that I follow," says Asprey, "and there is no way in heck I'm going to walk around ejaculating willy-nilly. I've got shit to do!"[12]

Asprey is constantly experimenting with smart drugs, but even these weren't enough to help him complete his book on time. To make the deadline, he had to pull five all-nighters in a row, while remaining alert and productive.

To achieve this, he consumed a combination of coffee, carbs, and medium-chain triglyceride (MTC) oil and attached electrodes to his temples that ran a small current through his brain for ninety minutes a night. This is called cerebral electrical stimulation. Asprey also took cortisol, held a violet laser over his head to help his brain mitochondria grow, and worked at a standing desk with a spiky mat to stimulate his feet.

To get even more blood to his brain, every half an hour he would go to a whole-body vibration plate and do handstands with his feet against the wall. After five days and nights, he turned in the manuscript on time and says his publisher was happy with it. Again, none of these methods have been proved scientifically. They are just one biohacker's experience, but it shows you just how far people are willing to go when experimenting on themselves.

Asprey isn't alone. Silicon Valley is teeming with technology-focused self-improvement gurus. Many of them are hawking their own brands of smart drugs, health supplements, and odd gadgets to boost cognition and energy.

"You can give people lemonade and tell them it's a cognitive enhancer, and they'll get perky," says Derek Lowe, a science blogger and expert on drug discovery.[13]

It's the placebo effect. If patients believe in a drug or treatment, they can actually experience beneficial effects, even though the pill or treatment has zero therapeutic value. With no FDA approval for most of the nootropics and health supplements on the market and little in the way of trials underway, there's no way consumers can know for sure. It often comes down to faith in the product and promoter. Unfortunately, clever marketers can take advantage of the lack of scientific evidence.

Practically everyone used to agree that taking multivitamins was good for you. However, researchers have now concluded that multivitamins don't reduce the risk for cancer, heart disease, cognitive decline, or an early death. This is based on dozens of extensive studies over decades. Even worse, vitamin E and beta-carotene supplements can be harmful. Despite the abundance of evidence, more than 50 percent of Americans still take multivitamins. It's a $12 billion industry.

"Pills are not a shortcut to better health and the prevention of chronic diseases," says Larry Appel, director of the Johns Hopkins Welch Center. "Other nutrition recommendations have much stronger evidence of benefits—eating a healthy diet, maintaining a healthy weight, and reducing the amount of saturated fat, trans fat, sodium, and sugar you eat."[14]

Sometimes people with good intentions get ahead of themselves. We all want to believe there's a magic pill we can pop that will cure all of our problems and make our lives better. I have to admit that I'm one of those people who loves to try new things, and despite my reservations, I have experimented with a variety of supplements over the years. A turning point came when I was running my own Silicon Valley start-up. I was under tremendous pressure and needed to reduce my stress. That's when I heard of kava kava.

Apparently, the people of Fiji and Tonga have used this natural herb for hundreds of years as a traditional medicine to relax the body and mind. So I started taking it and felt a little better. That is, until the FDA announced that kava kava had been linked to liver scarring, hepatitis, liver failure, and even death.

Fortunately, I suffered none of these negative effects. But this was my wake-up call and changed how I view the supplements market. It doesn't matter if it's a natural herb used for hundreds of years or a newly minted drug, taking anything that hasn't gone through extensive clinical trials poses significant risks to your health. That doesn't mean nootropics and other supplements are worthless. In the future, we'll come to understand how our bodies function at a much deeper level, and I'm sure some of these supplements may turn out to be the wonder drugs we've all been waiting for. However, in the meantime, do you want to be the lab rat?

There are plenty of people, including some of my closest friends, who don't mind experimenting on themselves regardless of the dangers. If they understand the risks, that's their choice. For my part, I will happily watch from the sidelines and take notes as these brave souls attempt to become the next superhumans.

LIFE EXTENSION: THE EPIC OF GILGAMESH

Speaking of superhumans, who doesn't want to live forever? One person is my brother. He thinks one life is more than enough for his fragile soul and looks forward to departing on schedule. Most people, however, would like to extend their lives well into the future. This inclination dates back to the earliest recorded myths, like the Epic of Gilgamesh. In this ancient Mesopotamia poem, the hero embarks on a perilous journey to discover the secret of eternal life. He never attains immortality, but you just might.

All around the world scientists and biohackers are experimenting with new ways to extend life. Even big companies are stepping up to this age-old challenge. Google launched Calico, a company with the mission of tackling aging and increasing health span using the latest science and technology. They brought in heavyweights to head up the company. Arthur Levinson, the founder and CEO, is also chairman of the board of Apple and served as CEO of Genentech. David Botstein, chief scientific officer, comes from Princeton University, where he was director of the Lewis-Sigler Institute.

"Illness and aging affect all our families," announced Larry Page, the cofounder of Google, at the launch of Calico. "With some longer term, moonshot thinking around healthcare and biotechnology, I believe we can improve millions of lives."[15]

Page isn't alone. Silicon Valley is full of billionaires who believe the world would be a better place if they stuck around a little longer. Larry Ellison, CEO and cofounder of Oracle, donated $370 million to aging research, according to the *New Yorker*.

"Death has never made any sense to me," says Ellison. "How can a person be there and then just vanish—just not be there?"[16]

Peter Thiel's Breakout Labs is funding radical science and bold ideas, like growing bones from stem cells, repairing age-related cellular damage, and figuring out ways to rapidly cool and preserve organs.

"The proposition that we can live forever is obvious. It doesn't violate the laws of physics, so we will achieve it," says Arram Sabeti, a tech entrepreneur and investor in the Longevity Fund.

Pierre Omidyar, founder of eBay, has donated millions to research why some people are able to bounce back from diseases.

"I have the idea that aging is plastic, that it's encoded. If something is encoded, you can crack the code," says Joon Yun, a health-care hedge fund manager and donor to longevity research. "If you can crack the code, you can hack the code!"

There is hard science to back up life extension. The Greenland shark can live up to five hundred years and never gets cancer. Even clams can live more than five hundred years. Tortoises can live to two hundred. Research on nematode worms showed that a relatively simple gene mutation can extend their life by a factor of ten. So, what's stopping us from doing the same for humans?

The molecular geneticist Jan Vijg sums it up nicely: "You can't just copy a single mechanism from the tortoise. We'd have to turn our genome over to the tortoise—and then we'd be a tortoise."

The prevailing view among longevity scientists is that aging is a natural product of evolution. The benefit to dying is that it makes way for a new generation with improved genes, thereby increasing the species' chances of long-term survival. Different species have different life spans because that's what worked best for them.

"If you just kept aging at the rate you age between twenty and thirty, you'd live to a thousand. At thirty, everything starts to change," says Eric Verdin, CEO of the Buck Institute for Research on Aging.[17] After thirty our risk of mortality doubles every seven years. Once we pass along our genes and raise our young, what happens afterward doesn't matter from an evolutionary standpoint. This is why the body isn't programmed to keep us alive longer.

Unity Biotech, a darling of Silicon Valley, believes it can address this evolutionary limitation by targeting senescent cells. These zombie cells linger and proliferate as we age, emitting substances that cause inflammation and turn other healthy cells senescent.

"Using a genetic trick to get rid of these senescent cells can significantly improve health and lifespan," says Ming Xu, who was part of a Mayo Clinic team that ran similar experiments on mice.[18]

Another promising approach is to extend our telomeres, which are repeating segments of DNA at the end of each chromosome. They become shorter each time a cell replicates, until they get too short, which is perceived by the cell as DNA damage and leads to cell death. Telomere shortening is thought to be responsible for many age-related diseases. Telomerase gene therapy in mice has delayed aging and increased longevity without causing cancer.

The science, however, isn't clear. Animals with long telomeres don't necessarily have long lives. The more we know about the body, the more we realize how little we know. "At the beginning, we thought it would be simple—a clock!" explains Gordon Lithgow, a leading researcher on extending the life of nematode worms. "But we've now found about five hundred and fifty genes in the worm that modulate life span. And I suspect that half of the twenty thousand genes in the worm's genome are somehow involved."[19]

One of the more colorful characters exploring longevity is George Church, a geneticist at Harvard. With his big, bushy beard and shock of gray hair, he looks like Charles Darwin. However, his ambition is to take evolution into his own hands. He has cofounded more than 35 start-ups, coauthored 527 scientific papers, and filed more than 50 patents. He's accomplished all of this while suffering from narcolepsy, a condition which causes him to suddenly drop off to sleep, sometimes in the middle of a conversation. Often, he'll wake up a short time later with a brilliant insight to share.

One of Church's projects is to extend human life. He culled forty-five gene variants from super centenarians, people who have lived to 110, as well as from yeast, worms, flies, and long-lived animals. He's now using gene therapy to add antiaging instructions to DNA. His lab has succeeded in making mice reverse nine diseases of aging, and now it will begin testing on dogs.

"Aging reversal is something that's been proven about eight different ways in animals," says Church, who not only wants to help people live longer but also foresees a time when we never get sick again. "We have a strategy by which we can make any cell or any organism resistant to all viruses by changing the genetic code. So, if you change that code enough, you now get something that is resistant to all viruses, including viruses you never characterized before."

In addition to pushing the limits of science, Church enjoys pushing the limits of his colleagues. He once went so far as to submit a peer-review to the journal *Science* written entirely in genetic code—not English. It was treated as a normal review, even though the editors could not understand it, and later became the basis of a new industry: DNA data storage. He's also said to have survived on nothing but laboratory nutrient broth for months.

With the relentless way Church pursues his interests, it's no surprise he completed two simultaneous undergraduate degrees at Duke University in just two years. In grad school, he became so obsessed with rendering 3D images of molecules through X-rays and math that he ended up living in the lab.

"I was extremely excited about the research I was doing," says Church. "And so, I would put in 100-plus hours a week on research and then pretty much didn't do anything else."[20]

Despite his helping to reduce the cost of whole-genome sequencing from billions of dollars to just thousands, Duke expelled him from its PhD program for missing too many classes. That's how he wound up at Harvard.

Gene therapies aren't the only game in town. Jesse Karmazin, founder of the start-up Ambrosia, has launched an antiaging service for wannabe vampires with a lot of cash to spare. He claims that infusions of blood plasma from young donors can reverse aging. His company charges $8,000 per liter per treatment.

"I want to be clear, at this point, it works. It reverses aging," says Karmazin. "I'm not really in the camp of saying this will provide immortality, but I think it comes pretty close."[21]

The FDA disagrees, stating that "there is no proven clinical benefit of the infusion of plasma from young donors."[22] After this announcement, the site stopped offering services, but surprisingly, it resumed business months later. Karmazin claims the FDA didn't contact him after it issued the warning, and it has taken no enforcement action against his company.

Maybe this is because in 2005, Tom Rando, a stem-cell biologist and neurologist at Stanford, announced that his lab conducted experiments where the blood from young mice succeeded in rejuvenating the livers and muscles of older mice. However, in this experiment, the young mice were placed in parabiosis for over five weeks with the older mice; it wasn't simply a one-off transfusion but rather exposure to a systemic environment. So, the jury is still out on whether Ambrosia's treatment will extend human life.

In an attempt to extend life without resorting to vampirism, Steve Horvath, a geneticist at the University of California, Los Angeles, developed a widely used epigenetic clock to evaluate a novel antiaging cocktail of three substances developed by Greg Fahy. Horvath showed that the volunteers shed two and a half years of their biological ages on average after twelve months of treatment, and their immune systems showed signs of substantial rejuvenation.

"I'd expected to see slowing down of the clock, but not a reversal," says Horvath. "Because we could follow the changes within each individual, and because the effect was so very strong in each of them, I am optimistic."[23]

If this sounds too good to be true, keep in mind the study only involved nine participants and lasted just one year. Horvath is now conducting a larger clinical trial in order to validate the original findings.

One of the biggest trends in antiaging is doctors prescribing drugs "off-label" for their possible benefits. Rapamycin, for example, has been approved by the FDA to coat coronary stents, prevent organ transplant rejection, and treat a rare lung disease. But now it's being used off-label as an antiaging drug. This is because researchers have shown that it can prolong the life span of mice, worms, and flies, as well as prevent age-related conditions in rodents, dogs, primates, and even humans.

Many people don't want to wait for a clinical trial or FDA approval that may never happen. Rapamycin is a generic drug, which means none of the big pharma companies has a reason to fund an expensive trial.

"There's no profit," says Matt Kaeberlein, a professor of pathology at the University of Washington medical school. "Without profit, there's no incentive."[24]

Metformin is another popular off-label drug that shows potential. It's typically prescribed for diabetes and only costs about five cents a pill. Researchers found that diabetics taking metformin tended to live longer, had fewer cardiovascular events, and were less likely to suffer from dementia and cancer.

"Metformin may have already saved more people from cancer deaths than any drug in history," says James Watson, winner of the Nobel Prize for the discovery of DNA.[25]

Nir Barzilai, director of the Institute for Aging Research at the Albert Einstein College of Medicine, is one of metformin's biggest proponents and sees it as a way to increase health span. He believes having people live disease-free into their nineties and beyond is more important than having them live longer but poor-quality lives. Without the support of big pharma, Barzilai had to turn to the government and private donors in order to raise enough money to conducts clinical trials.

Libella Gene Therapeutics took another approach. The company circumvented FDA approval entirely by heading to Columbia. They are offering wealthy patients the opportunity to participate in their telomere-lengthening clinical trial at $1 million per dose. Is this unethical, or is it a good way to speed up and fund human trials?

"It takes a lot longer, is a lot more expensive, to get anything done in the US in a timely fashion," says Jeff Mathis, president of Libella.[26] He feels they are doing their patients a service by making this option available at $1 million a pop.

Jerry Shay, an expert on cancer and aging at the University of Texas Southwestern, disagrees. He thinks the dangers are significant, pointing out that there's a significant risk of activating precancerous cells.

If you'd like a less risky way to extend your life, you may want to try caloric restriction. It won't cost you millions and will save you money on grocery bills. Jack Dorsey, the cofounder and CEO of both Twitter and Square, eats only one meal on weekdays. On weekends, he tries not to eat at all. He also starts off each morning with an ice bath before walking five miles to the Twitter headquarters. Dorsey also drinks salt juice and uses a near-infrared sauna tent that purports to block electromagnetic transmissions. To cap it off, he wears an Oura Ring in order to track his activity and sleep. Geeks need their data.

Probably one of the most famous antiaging biohackers in Silicon Valley is Ray Kurzweil. He believes that if we can just stay alive long enough for science to solve the problems of aging, we can all have a shot at living forever. He was born in 1948, and in his quest to achieve immortality he has tried all sorts of approaches, including regular intravenous longevity treatments. He's also willing to spend "a few thousand dollars per day" on food, supplements, and pills.[27] This comes to about $1 million a year. He even hired a pill wrangler to take his typical allotment of a hundred pills out of their bottles and sort them into daily doses.

According to market-research firm Global Industry Analysts, the antiaging industry generates more than $80 billion per year. And with all this money comes a downside. There are a lot of snake oil salesmen out there, who are happy to cash in on anyone willing to believe in the fountain of youth.

"Much of the aging field is charlatans," says Barzilai. "They tell you take this or that and you'll live forever. But you have to do a clinical trial that is placebo-controlled, and only then can you say what it really is and whether it's safe."[28]

One of the most hyped and heavily marketed antiaging supplement companies is Elysium. With Leonard Guarente, director of aging research at MIT, as its cofounder and chief scientist, and six Nobel laureates as advisers, it seems like a sure bet. Even better, one of the main ingredients in its supplement is a chemical called nicotinamide riboside, which has shown promise in making mice live longer.

Unfortunately, mice aren't people. What works in a mouse doesn't necessarily translate to human beings. Mice don't have heart attacks, and their muscles start wasting suddenly, rather than gradually as in humans. There are thousands of drugs that have showed promising results in mice, rats, and other animals, including dozens of cures for cancer, but when it came to people, they wound up duds, or even worse, had harmful effects. In the end, it doesn't matter how many Nobel laureates a company signs on; the proof is in the human trials, and until those are completed, it's anyone's guess.

The problem is that it's difficult to know whom to trust and even harder to find reliable data. Running trials for cutting-edge treatments is extremely expensive and takes a long time, especially when it comes to aging. Many companies simply don't want to make the investment. Realistically, it may be years, or even decades, before we know what works and what doesn't, and for a lot of elderly people, that may be too late. That's why there's a temptation to take a chance on a supplement or treatment, even when the results are unknown.

THE GLOBAL IMPACT OF LIFE EXTENSION

Many people worry about what happens if humans can live to be 150 or more. Will this severely impact everything from food production and housing to the labor market?

Before we delve into that, let's put things in perspective: Accidents and violence, not age-related diseases, are the leading causes of death up to age forty-four. After this, cancer tops the list, and at sixty-five, it's heart disease. Even if we find a cure for cancer, humans would only increase their average life span by 3.3 years. Curing heart disease would secure us an extra four years. If we could eliminate all diseases, the average age would stretch into the nineties, but to get beyond this, we will have to slow or reverse aging. There's no other path forward.

What's clear is that despite all the missteps and complications, researchers are making progress, and we may be nearer to a breakthrough than some anticipate. How close are we? No one knows. But there's a pretty good chance we will see many diseases, like cancer, go from life threatening to manageable conditions in the next couple of decades. At the same time, our life expectancy will continue to rise, especially for those with the income and education to take advantage of cutting-edge treatments.

Will we be able to live forever? Obviously not. Nothing lasts forever, not even our universe. But we certainly will be able to extend our lives considerably. So, the real question is not if but when. And when it does happen, what does that mean for us? How do we manage a world where people can live hundreds of years? Will our planet become overpopulated? Will we be forced to ship people off to Mars and other planets just to make room on Earth? Maybe not.

As we've seen, populations tend to stabilize and even decline as the standard of living rises. Just look at Japan. It's set to lose the equivalent of a midsize city every year for the foreseeable future. We see similar trends in Korea, China, and Europe. The idea that people may live longer, healthier lives into their eighties and beyond may be a good thing.

In the future, most people in industrialized countries with access to longevity treatments may only choose to have one or two children in their extended lifetimes. And no matter how advanced our technology, people will still die in vehicle accidents, in natural disasters, and from diseases, so things may balance out at some point.

If the world's population continues to rise, it shouldn't be a problem. We still have a lot of capacity to absorb more people. With advances in food production, housing, medical care, renewable energy, and transportation, humanity will find a way to manage. We are a remarkably ingenious and adaptable species.

On the other hand, if the world's population follows that of Japan and declines too rapidly, we may have to give up relying upon natural births and begin producing test-tube babies using artificial wombs—and if that doesn't work, cloning is always an option. Either way, I'm not too worried and look forward to living well into my hundreds.

CRYOGENICS, REBIRTH, AND RABBIT BRAINS

It's one thing to live forever and another to bring people back from the dead. Jesus managed to pull it off, but can a company or foundation, no matter how well funded, resurrect the dead? I have my doubts, but there are a lot of believers out there.

Walt Disney may have been one of them. I grew up hearing about how Disney's corpse was cryopreserved, but this turns out to be a myth. The rumor probably got started in 1972 when Bob Nelson, who was the president of the Cryonics Society of California at the time, told the *Los Angeles Times* that Disney wanted to be frozen.

"The truth is, Walt missed out," clarified Nelson. "He never specified it in writing, and when he died, the family didn't go for it."[29]

In January 1967 Nelson froze the first man, Robert Bedford, who died only one month after Disney did. This may have contributed to the Disney myth.

What's surprising is that Nelson didn't have a scientific background. He didn't even graduate from high school. Yet he had become the de facto spokesperson for this fledging industry. Unfortunately, years later, Nelson's endeavor ran out of money, and he simply locked the vault and left. It was only when the media publicized this that family members found out their loved ones were decomposing. They filed a lawsuit against Nelson and won a judgment of $800,000. Nelson responded by writing the book *Freezing People Is (Not) Easy*, revealing his side of the story.

There are now a number of foundations and companies dedicated to developing and selling cryonics as a service. The tantalizing idea behind cryonics is that if someone has died from a disease that is incurable now, this person can be frozen and then revived in the future when a cure has been discovered. Today, many people believe aging is a curable disease, so practically anyone with a mostly intact body upon death would qualify.

Keep in mind that cryonics isn't cheap. It typically costs up to $200,000 to have your whole body preserved. If you happen to be a more frugal futurist, you can spend much less to preserve just your brain in perpetuity. This option is known as neuro-suspension. However, you must continue to pay an annual maintenance fee for the brain, usually around $400, or you will be unplugged. So, you need to be sure to set up a fund for this, too.

Here's how it works. Assuming you pass away in a hospital with a cryonics team at your bedside, they will spring into action as soon as you are pronounced legally dead. The team will stabilize your body, supplying your brain with enough oxygen and blood to preserve minimal function until you can be transported to the suspension facility.

At the facility, the actual freezing begins. First, they remove all the water from your cells and replace it with a sort of human antifreeze. This is to protect the organs and tissues from forming ice crystals. Then they place your body on a bed of dry ice until it reaches minus 130 degrees Celsius. Finally, they place your body in one of those sci-fi-looking metal tanks filled with liquid nitrogen and cool you to around minus 196 degrees Celsius. You'll be stored head down, so if there is ever a leak, your brain will remain immersed long enough for them to plug the hole.

Does this sound fun? When I was a kid, I thought so. I swore that if I had the money, I would do it. I'm not the only one. Numbers are hard to come by, but today, according to the *Journal of Medical Ethics*, more than 250 bodies have been cryopreserved in the United States, and 1,500 people have made arrangements for cryopreservation when they pass away.

"I prefer cryonics to uploading," says Peter Thiel. "If cryonics works, you can still be the same person. If you were uploaded onto a computer and you had all your information represented, it's not clear whether that's genuine immortality or not."[30]

Two more Silicon Valley luminaries, Ray Kurzweil and Aubrey de Grey, chief science officer of the SENS Research Foundation, are believers and see cryonics as a good backup plan if they don't live long enough to realize immortality. The real question is, Does it work? "Human embryos are routinely preserved for years at temperatures that completely stop the chemistry of life," states the Alcor Life Extension Foundation. "Adult humans have survived cooling to temperatures that stop the heart, brain, and all other organs from functioning for up to an hour."[31]

Scientists have managed to cryogenically freeze a rabbit brain and recover it in an excellent state. This, however, does not mean the poor rabbit's memories or sense of self can be recovered. No one has proven this yet.

"The features of your neurons (and other cells) and synapses that make you 'you' are not generic," points out Michael Hendricks, a neuroscientist and assistant professor of biology at McGill University. "While it might be theoretically possible to preserve these features in dead tissue, that certainly is not happening now. The technology to do so, let alone the ability to read this information back out of such a specimen, does not yet exist, even in principle."[32]

Personally, I've given up my childhood dream of being frozen. For anyone thinking of being frozen today, it's highly unlikely that even the most advanced future society will be able to restore "you" as you are right now. Even if they bring you back, you may be trapped inside a broken body, both mentally and physically damaged and perhaps suffering chronic pain, as well as some serious brain damage. Do you want to run this risk? To me, a conventional death sounds preferable.

THE CLONES ARE COMING

If freezing your body sounds bizarre, what about cloning yourself?

Research into cloning dates back to 1885, when Hans Driesch, a German biologist and philosopher, demonstrated that by merely shaking a two-celled sea urchin embryo he could separate the cells, which would each grow into a sea urchin. In 1938 Hans Spemann, another German scientist, then proposed a "fantastic experiment" to replace the nucleus of one egg cell with the nucleus of another cell and grow an embryo from such an egg. Jump ahead to the 1990s and along comes Dolly the sheep, the first mammal cloned from an adult somatic cell, using the process of nuclear transfer.

Today, animals are being cloned all over the world. Livestock are leading the way. Pollard Farms looks like any other ranch, except that some of the cattle are identical copies of one another. "We're trying to stay on the very top of the heap of quality, genetically, with animals that will gain well and fatten well, produce well and reproduce well," says owner Barry Pollard.[33]

In 2008, the FDA approved the sale of cloned meat. Ranchers are now using cloning to breed cows that produce more and tastier milk and bigger, juicier steaks. Cows can also be selected for their health, resistance to disease, and ability to reproduce with clockwork precision. As the practice of cloning matures, we will see more farms around the world filled with perfect cattle, sheep, pigs, and other livestock.

The problem comes when diseases mutate and livestock that was once disease resistant becomes vulnerable. A lack of genetic diversity means a single bacteria or virus that might have wiped out a small percentage of a herd may wind up killing every single animal. As the cost of cloning comes down and the practice gains widespread adoption, we may wind up with very little genetic diversity from ranch to ranch and even country to country. This could jeopardize the worldwide supply of livestock.

Unlike GMOs, clones are not genetically modified. They are copies of the original animal. But does that mean they are safe for humans to eat?

"We don't think that cloning is a technology that's ready yet, and we certainly don't think it's ready to be on your plate," warns Jaydee Hanson, director of policy at the Center for Food Safety, who believes not enough testing has been done yet.[34]

Whatever you may feel, cloned meat is still too expensive for the dinner table, and it will be a while before the costs come down. In the meantime, cloning is finding other applications. Some dogs are exceptionally gifted at sniffing out drugs, bombs, and other illegal substances. Breeding these dogs can be time consuming and expensive. That's why the Beijing police force welcomed six cloned dogs into its ranks.

Adolfo Cambiaso, one of the world's best polo players, has embraced cloned horses. Riders must frequently change mounts during a polo match and having a physiologically identical horse can make the transition easier. The riders don't have to change their riding style midgame. A cloned horse costs around $120,000. But there's plenty of money in professional polo, and the riders will spend what it takes to win.

Even more interesting is when a rider becomes emotionally attached to a particular horse. They don't want to lose it, yet most professional polo ponies need to be retired after the age of ten. "Seeing Cura alive again after so many years was really strange," says Cambiaso, who cloned his horse after its death. "It's still strange. Thank goodness I saved his cells."[35]

Many people think of their dogs and cats as part of the family and the thought of losing them someday can be as traumatic as losing a son or daughter. "I was so devastated by the loss of my dear Samantha, after 14 years together, that I just wanted to keep her with me in some way," says Barbra Streisand, the famous singer, actress, and director. "It was easier to let Sammie go if I knew that I could keep some part of her alive, something that came from her DNA."[36] Streisand had Sammie cloned, not once but twice. The results are Miss Violet and Miss Scarlett.

Sooam Bioengineering Research Institute, the world's largest animal cloning factory, has cloned more than one thousand dogs at up to $100,000 per birth. "Yes, cloning has become a business," says Hwang Woo-suk, a South Korean veterinarian and researcher at Sooam Biotech. "If the cells from the dead dog are not compromised, we guarantee you will get a dog within five months."[37]

To produce these clones, they require surrogate mothers. "Surrogate mothers are a little bit like *The Handmaid's Tale*," said Jessica Pierce, an ethicist at University of Colorado. "It's a canine version of reproductive machines."[38]

Hwang has already run into trouble with the Korean government. He was charged with bioethics law violations and embezzlement when it emerged that some of his stem cell research had been faked. Despite the controversy, Hwang's lab continues to actively publish manuscripts and expand operations. The Chinese biotech company Boyalife Group announced that it will partner with Sooam Biotech to open an even bigger cloning factory. Their goal is to produce up to one million cattle embryos per year to meet growing demand for quality beef in China.

If cloning cattle, cats, and dogs is troublesome, what about humans? That's where the issue gets really controversial. Imagine a parent who lost a child in an accident or to disease. The temptation to bring a dead child back to life would be enormous. How much would a parent pay? And whom does it harm?

Fortunately, we have some time to figure this out. Cloning humans and primates is much harder than dogs and cats. But that doesn't mean it's impossible. Scientists in China have already succeeded in cloning two macaque monkeys, and more experiments are underway.

The market for human clones is broader than you may think. Millions of couples are infertile, and many of these may opt to create children out of their DNA, if it were legal. Is the perfect nuclear family of 2050 a married couple with the son being the exact copy of the father and the daughter being a copy of the mother? Would this take us a step closer to immortality? Or is it so disturbing we should never go there?

There are broader social issues that come into play. Cloning is expensive and may remain so for quite some time. Is this only for rich folks, like

Barbara Streisand? And what would this do to the human gene pool several generations down the line?

What we think of as bizarre and disturbing now may not be so unthinkable as the technology matures. Social norms change with technology. Just look at in vitro fertilization (IVF). In 1978 Louise Brown, the first IVF baby, was born. Today, there are millions of IVF babies, and people don't think it's strange at all. Most of us forget how controversial it was.

"It was viewed with absolute suspicion," says Peter Braude, head of the Department of Women's Health at King's College London. "If you talk to people today about human reproductive cloning, the feeling you get, that it is playing God, is just how it was in 1978 with IVF."[39]

We can think whatever we like now, but in the future, having a clone of yourself may be considered normal. Keep in mind that your clone wouldn't be you. Personality is shaped not only by heredity but also by environment.

Streisand realizes that her puppies aren't exact copies of Sammie. "They have different personalities," she says.[40]

Is it even possible to preserve pets' memories and personality? Sinogene Biotechnology thinks so. Their plan is to use artificial intelligence and a brain-machine interface to store dogs' memories and then pass them along to the clones.

In the future, will humans do the same? Would you consider growing a clone of yourself and then using an advanced brain-computer interface to transfer your memories to this younger, healthier version? If you did this, would the clone be you? Or would it be someone else? What if the original, older version of you is still alive? In this case, you'd have two of you, each claiming to be the real thing. That may be confusing for your loved ones.

If transferring memories isn't possible, what about genetically engineering your clone to be born without a brain and then transplanting your actual brain to the clone? This would avoid the problem of having duplicate selves. If a brain transplant won't work, what about surgically removing your head and attaching it to the cloned body? This sounds grotesque, but there are scientists who are working on it.

"The first human transplant on human cadavers has been done," says Italian surgeon Sergio Canavero. "A full head swap between brain dead organ donors is the next stage."

If this isn't enough to remind you of Dr. Frankenstein, Canavero said that an operation on a live human being is imminent. His critics not only condemned the experiment as reckless, but they pointed out that just because a patient may survive doesn't mean the person's thoughts and personality will remain intact.

"The person will encounter huge difficulties to incorporate the new body in its already existing body schema and body image," Italian scientists wrote in *Surgical Neurology International*. The report cited similar problems in face and hand transplants that resulted in serious psychological issues, "namely insanity and finally death."

Canavero remains undeterred. "I understand humans love the gory side of the surgery, but this is a medical procedure for a medical condition for people who are suffering awfully," retorts Canavero. "So, it's not a joke."[41]

All this may seem crazy, but researchers at the Yale School of Medicine have succeeded in keeping the brains of decapitated pigs alive for as long as 36 hours.

"It's at the extreme of technical know-how, but not that different from preserving a kidney," says Steve Hyman, director at the Broad Institute of Harvard and MIT. "It may come to the point that instead of people saying, 'Freeze my brain,' they say 'Hook me up and find me a body.'"[42]

That body may be your clone in cold storage waiting for the chance to serve you. How's that for a bright future?

THE BIONIC BODY

In addition to creating a clone of yourself, someday you might be able to repair and upgrade your existing body with the use of bionics. I grew up watching all the cyborg flicks, from *The Six Million Dollar Man* to *RoboCop*. Now it seems like my childhood fantasies are coming true. We are at the point where bionics are a reality.

Kerry Finn lost his left leg from vascular disease brought on by type 2 diabetes. Fortunately, he was one of the first to test out the University of

Utah's bionic legs. Unlike a typical prosthetic leg, Finn's new bionic leg is self-powered and autonomous. Containing a microprocessor and motorized joints, it automatically reacts to how your body moves in much the same way a natural limb does.

"If you've ever seen *The Terminator*, that's what it was like," says Finn. "It made me feel like I could do things I could not do before. Every time I made a step, it was an awesome feeling."[43]

Finn can't run at sixty miles per hour like Steve Austin, the fictional hero of *The Six Million Dollar Man*, but the bionic leg does make life easier.

"If you walk faster, it will walk faster for you and give you more energy," explains Tommaso Lenzi, the project lead.[44] The leg is exceptionally lightweight and powerful. It's also smart. It can adapt to the wearer's movements without the person having to consciously think about it.

At Johns Hopkins Applied Physics Lab, they are developing a robotic arm that the wearer can control just by thinking. It's great for people like Nathan Copeland, who was paralyzed from the chest down in a car accident and cannot feel or move his hands.

"We wired the robotic arm to Nathan's brain, providing him with two-way electrical feedback," says Michael McLoughlin, the chief engineer on the project. "He not only could operate the device by thinking, he also received signals coming from the robotic arm, such as the feeling that his fingers had been touched. It was a real *Star Wars* moment."[45]

"I can feel just about every finger," says Copeland. "It's a really weird sensation. Sometimes it feels electrical, and sometimes it's pressure, but for the most part, I can tell most of the fingers with definite precision. It feels like my fingers are getting touched or pushed."[46]

At the University of Utah, Dr. Gregory Clark's team has developed the LUKE Arm, named after Luke Skywalker. This advanced prosthetic enables amputees to feel the sensation of touch via an array of electrodes implanted in their nerves. Keven Walgamott lost his left hand and part of his arm in an electrical accident.

"It almost put me to tears," said Walgamott, when he first experienced the LUKE Arm. "It was really amazing. I never thought I would be able to feel in that hand again."[47]

In the near future, you won't need to wear a prosthetic to become a bionic man or woman. A new breed of exoskeletons is here to give you Incredible Hulk–like powers. The US Army is developing one right now. It's called FORTIS and uses AI to analyze and enhance a soldier's movements. Powered by lithium ion batteries, the robotic exoskeleton bears most of the weight, allowing soldiers to carry 180 pounds up five flights of stairs with far less energy expenditure.

"We've had this on some of the Army's elite forces, and they were able to run with high agility, carrying full loads," says Keith Maxwell, senior program manager at Lockheed Martin, the company behind FORTIS.[48]

Exoskeletons aren't just for the military. US Bionics has developed a series of industrial exoskeletons designed to help workers handle large loads without straining their backs, arms, and legs. The goal is to reduce job-related injuries and make workers more productive.

"These exoskeleton systems for workers basically minimize the stress and strain in some particular joints," says Homayoon Kazerooni, the CEO of US Bionics and a professor at University of California, Berkeley. "We're looking at bionic devices as affordable consumer products, such that workers can buy them."[49]

In the future, will all of us be walking around in Iron Man suits? Probably not, unless we happen to be living on Mars or some other planet. But affordable exoskeletons will dramatically improve some people's lives. Those who are paralyzed or suffer from mobility problems may someday opt for an exoskeleton over a wheelchair.

Paralyzed from the waist down after a BMX bike accident, Steven Sanchez traveled around the world wearing an exoskeleton, and it became a necessity for him. Adam Gorlitsky, who is also paralyzed from the waist down, completed the 2020 Charleston Marathon in just 33 hours, 50 minutes, and 23 seconds with the help of an exoskeleton.

Keep in mind that exoskeletons are still in their infancy. The price, weight, battery life, and form factor will improve dramatically over the coming years. Someday, people may wear these exoskeletons underneath their clothing and walk around without anyone realizing it. So, if you're backpacking in the mountains, clearing out your garage, or just feeling exhausted, you may don a super lightweight, flexible pair of exopants and go about your day without a second thought.

THE CRISPR REVOLUTION

In the 1950s, when Crick, Watson, Wilkins, and Franklin discovered the double helix structure of DNA, they gave birth to modern genomics. They revealed how it was possible for genetic instructions to be held inside organisms and passed from generation to generation. This breakthrough set the stage for the rapid advances in molecular biology and genetic engineering that we are seeing today.

Now that we have invented advanced gene-editing technologies, like CRISPR, the door is wide open for developing all types of new therapies, diagnostic techniques, and genetically modified organisms. These have the potential to radically improve our lives and the world, as well as unleash forces that may forever change life on this planet in ways we cannot predict.

Let's begin with the food we eat. Back in 1994, Calgene brought the first genetically engineered crop to market: the Flavr Savr tomato.

"The tomato stays riper, longer than the nonengineered variety, and they say it's tastier," announced Tom Brokaw on NBC Nightly News.[50]

The backlash was immediate. Most people weren't ready for genetically modified foods. Critics labeled it "Frankenfood," and the Flavr Savr was eventually shelved. But that didn't stop companies like Monsanto, which acquired Calgene, from continuing development of new genetically modified crops. Today, in the United States, 93 percent of soybeans and 88 percent of corn are genetically altered. Whether you like it or not, more than 60 percent of all processed foods in US supermarkets, including cookies, ice cream, pizza, chips, salad dressing, corn syrup, and baking powder, contain ingredients from genetically modified soybeans, canola, or corn. Unfortunately, none of these processed foods are labeled as containing genetically modified organisms (GMOs). This is a testament to the power of corporate lobbyists in America.

In Europe it's a different story. The majority of member states of the European Union have gone to the other extreme, voting to ban GMOs partially or fully. They are afraid of the long-term health consequences. The fact is that no one really knows what will happen as we begin experimenting on our food supply. So far, GMOs have been found to exhibit no toxicity, even across multiple generations. But does this mean they're safe?

The answer is that we cannot know for certain, and we may not have the luxury of waiting to find out.

Even if you buy organic foods, you may not be able to avoid genetically modified foods. GMO crops can contaminate organic crops through cross-pollination in the fields, as well as seed and grain mixing. Once you release a genetically modified organism into the wild, it's extremely hard to control. This has created a real headache for organic farmers, who have been in many pitched disputes and legal battles with Monsanto over everything from genetic pollution to patent rights.

"We found transgenic plants growing in the middle of nowhere, far from fields," says ecologist Cynthia Sagers, a professor at Arizona State University. "Not only is it going to jump out of cultivation; there are sexually compatible weeds all over North America."[51] In other words, we may be in the process of creating superweeds that are resistant to even the toughest pesticides.

Despite all the problems associated with GMO crops, the benefits are real. Let's take the climate crisis we are now facing. Not only will climate change cause more extreme weather conditions, but it will increase the impact of crop-destroying pests and diseases. Pairwise, a Monsanto-funded start-up, is taking on this challenge. They are using CRISPR gene-editing technology to develop new strains of crops able to resist disease, tolerate droughts, and survive floods, as well as taste better and last longer on store shelves.

Genetically modifying crops is not new. Humans have been changing the genetic makeup of plants ever since agriculture began. Selecting seeds from the best crops and breeding and crossbreeding varieties is part of our history. Almost none of the vegetables and fruits we eat today are what grew in the wild thousands of years ago. They tend to be sweeter, larger, and longer-lasting. They also have much higher yields. The only problem is that it takes time, lots of time, to modify plants through traditional breeding techniques. Through modifying and editing a plant's genes, we can speed up the process and gain far greater control over the outcome, which may wind up being vital for our survival in a world with less water, more extreme temperatures, and more virulent pests and diseases.

Fifty years from now, we may come to consider genetically edited and modified crops not only safer but preferable to organic foods. Imagine a world where we have wheat, soy, and peanuts that are allergen-free, bananas that deliver vaccines, and vegetable oils loaded with therapeutic ingredients for reducing the risks of cancer and heart disease. In our grocery stores, we may find new, exotic fruits and vegetables, like mangonanas, bluepples, and carroccoli. We may even have precision crops that are modified to suit our individual genetic and dietary needs. Instead of taking drugs, we may consume genetically altered produce to avoid conditions like diabetes, high cholesterol, and high blood pressure.

Even better, our foods will be tastier and healthier. Desserts could turn into health foods, similar to Woody Allen's classic movie *Sleeper*. You may bite into a dessert banana that tastes like caramel. Anyone up for a box of Krispy Kreme donuts that's better for you than a plate of brussels sprouts? I am! But only if we can ensure that genetically tampering with our food supply is safe.

"Every transgenic organism brings with it a different set of potential risks and benefits," says Allison Snow, a plant ecologist at Ohio State University. "Each needs to be evaluated on a case-by-case basis. But right now, only one percent of USDA biotech research money goes to risk assessment."[52]

It's not just crops we need to consider. It's also livestock. We are on a path to genetically modifying the meat we eat. The University of Florida is now working on using CRISPR to breed heat-resistant cattle that can survive in a world with more extreme temperatures.

In China, researchers are using CRISPR to create low-fat pigs. They inserted a mouse gene into embryonic pig cells and then proceeded to coax those cells to generate more than two thousand pig embryo clones. Female pig surrogates gestated the embryos, giving birth to twelve male piglets. As they matured, these pigs had 24 percent less body fat than pigs without the gene.

This is just one of many experiments going on. In the future, we'll see genetically modified low-fat pork, beef, and chicken in our grocery stores. We'll also see bigger, faster-growing animals that save farmers money and time. Already, AquaBounty, a Massachusetts-based start-up, is developing genetically modified salmon that grows twice as fast.

"It's identical to Atlantic salmon, with the exception of one gene," says Sylvia Wulf, CEO of AquaBounty.[53] The company is already selling its salmon in Canada.

If you want to really let your imagination run wild, there may come a time when we are eating animals that never existed before. If you modify enough genes, when does an animal become a new species? And once it's a new species, what's to stop us from going further? A cow a hundred years from now may not resemble a cow at all. It may have no legs, no head, and float in a vat of water in a factory farm. If this sounds grotesque, just look into how factory farms operate today, and you'll see it's not too off the mark.

Researchers may decide to cross a pig with a chicken to produce a picken. Anyone up for a slice of porken? The farms of the future could be filled with hybrid animals, whose meat offers foodies a huge variety of flavors, textures, and health benefits. If it seems hard to believe that people will actually eat something as unnatural as genetically altered, hybrid meats, just look at the processed foods that line our store shelves today. Have you read the ingredients? Do you understand half of what's in these products? Yet, people eat this stuff regularly. Why? Because it's engineered to taste delicious, and that makes it hard to resist. Like it or not, this trend is bound to continue and accelerate as new genetically altered foods enter the marketplace.

Another way CRISPR may affect the livestock industry is by helping to curb the rise of antibiotic-resistant bacteria. Not only are doctors overprescribing antibiotics to their human patients, but farmers are using huge amounts of it in animal feeds. The result is that we are breeding superbacteria at an alarming rate. It's estimated that more than 2.8 million people are infected with antibiotic-resistant bacteria each year in the United States alone, and more than thirty-five thousand people die. Health experts predict that threats from antibiotic resistance could drastically increase in the coming decades, leading to some ten million drug-resistant disease deaths per year by 2050.

In an effort to avoid having to give livestock antibiotics, Northwest A&F University in China is using CRISPR to create tuberculosis-resistant cattle. If this is successful, the hope is to develop pigs, chickens, ducks, and other livestock that can resist a wide variety of common diseases.

Locus Biosciences is taking another approach in the battle against antibiotic-resistant bacteria. The start-up is using CRISPR to create therapeutics that could eventually replace antibiotics altogether. All the antibiotics brought to market over the past 30 years have been variations on existing drugs discovered by 1984. This is because it's a challenge to develop substances that can kill bacteria and are nontoxic to humans. It's also not very profitable, which is why the major pharmaceutical companies have steadily cut back on investment in this area. Locus Biosciences said their first drug will target multidrug-resistant *E. coli* strains and can be developed in a profitable manner. If this proves to be true, we could see other CRISPR-generated antibiotics come to market.

The University of California, San Diego, is going directly to the source. They are developing a CRISPR-based system that inactivates antibiotic-resistant genes right inside bacteria. If it works, they may be able to eliminate drug-resistant microbes before they can infect livestock and humans.

The potential for CRISPR to benefit humanity is enormous, and this is why, despite the risks, we can't simply ban gene editing and genetically modified organisms. That would be a big mistake. These technologies can allow us to address many of the most urgent problems we face. The question shouldn't be whether or not to use this technology but how to use it responsibly.

GENE DRIVES

Another application of CRISPR is gene drives. A gene drive is a genetic modification designed to spread through a population at higher-than-normal rates of inheritance. Using CRISPR, scientists can modify certain genes in a plant or animal and then guarantee that these genes are passed down to subsequent generations, even if only one parent has those genes. By inserting the chosen gene into both strands of the offspring's DNA, a gene drive can make a recessive trait function as a dominant one.

Today, implementing a gene drive is not only possible but a viable way to fight invasive species. Take New Zealand, for example, where invasive rats are devastating the native bird populations. Every year, rats gobble up more than twenty-six million chicks and eggs. New Zealanders have been struggling to rid themselves of the rats using poison, guns, and traps.

They've had some success, but the rat populations keep rebounding, while the bird populations continue to decline.

To help combat the problem, Kevin Esvelt, a professor at MIT Media Lab, traveled to New Zealand and proposed a gene drive that could wipe out the invasive rats. This would save millions of dollars and be a permanent solution. To conservationists, this would be a godsend.

"Something is going to die," says James Russell, an ecologist born and raised in New Zealand. "Either a bird is going to be killed by a rat that we brought here, or we're going to kill the rat. And I would rather humanely kill the rat than have the rat inhumanely kill a bird."[54]

Many locals, especially Christians and Māori, don't agree. They feel that tampering with the process of natural selection is unethical. They also believe gene drives may wind up causing more harm than good. What if the genetically modified rats don't stay confined to the islands of New Zealand? Rats have a habit of jumping on and off ships. They may wind up spreading their genes to other rat populations around the world. Could this small experiment eventually drive rats to extinction globally?

Esvelt, who originally championed the idea, now has reservations: "It was profoundly wrong of me to even suggest it because I badly misled many conservationists, who are desperately in need of hope. It was an embarrassing mistake."[55] He further says, "The kind of gene drive that is invasive and self-propagating is in many ways the equivalent of an invasive species."[56]

After conducting mathematical simulations with colleagues at Harvard, Esvelt concluded that gene drives may be scarier than the problems they are intended to solve. Even the weakest gene drives, if not carefully designed, could destroy an entire species.

Russell agrees with being careful but isn't ready to dismiss gene drives as a solution, noting that more than 60 percent of the vertebrates that have disappeared from the planet have disappeared from islands, and in half of those cases, invasive species are to blame. He thinks we're smart enough to figure out ways to control the gene drives without causing an entire species to go extinct.

One solution may be gene drives that aren't designed to kill animals or plants. Esvelt proposed running a trial on the island of Nantucket to see if they could eliminate Lyme disease by modifying the genes of the mice that spread it. Instead of wiping out the mice, he'd use the gene drive to vaccinate mice against Lyme disease from birth. Would this be an acceptable risk? Esvelt is conflicted on the subject. He sees the value in gene drives for controlling the spread of disease but feels that we still don't understand the long-term consequences.

There are certain species that most of us wouldn't mind disappearing from the planet. One of them is mosquitoes. They've spread misery ever since the dawn of humanity. Even in the modern age, malaria is one of the world's most deadly diseases. Each year, an estimated three hundred to six hundred million people suffer from malaria. It kills more than one million people a year, most of them children. Even more worrisome, in many parts of Africa, 70 percent of malaria cases are resistant to inexpensive antimalarial drugs.

Andrea Crisanti, a geneticist at Imperial College London, has been working on creating a gene drive that could potentially wipe out the type of mosquitoes that transmit malaria. His genetically modified female mosquitoes cannot bite and do not lay eggs. In the lab Crisanti demonstrated that within eight to twelve generations, the caged populations produced no eggs at all. Target Malaria, a nonprofit, wants to bring this technology to Africa and other malaria hot zones.

Researchers from Johns Hopkins University are taking a different approach. Instead of annihilating mosquito populations, which may take a heavy toll on the animals that feed upon them, they are working to engineer mosquitoes that are resistant to the malaria parasite. By deleting a gene that enables malaria to survive in the mosquito's gut, they could stop infections, while still preserving mosquitoes.

This sounds promising, but even these gene drives come with certain risks. The danger is that a failed gene drive may ultimately make the mosquitos more resilient by expanding and modifying their gene pool. It's too early to tell how great this risk actually is, but it's something scientists are carefully monitoring.

DARPA, an agency of the US Department of Defense, has been funding a lot of the research into gene drives. In fact, Esvelt took money from DARPA for his own projects. Conspiracy theorists can have a field day with this, claiming the United States is preparing to weaponize gene drives or engage in eugenics. But the reality is much simpler. Gene drives are a threat to national security, and the military needs to prepare for it.

What if terrorists get ahold of gene drive technology? Could they unleash an attack on the world's food supply? Could a neo-Nazi group develop a gene drive that targets specific races, thereby instigating a genetic holocaust? The possibilities are truly horrifying, and the scary part is that the technology is readily available today. It doesn't cost much to set up a lab and begin experimenting. It may not matter if people have good or bad intentions; the technology is so powerful and so unpredictable that if we don't figure out ways to control or limit it, it has potential to do irreparable harm.

The fact is that gene drive technology has arrived, and we must deal with it. Just like with atomic energy, we have unlocked one of nature's great secrets. The question is whether we can figure out how to use this technology for the betterment of the world without permanently harming our environment and ourselves in the process.

COW CULTURES: LAB-GROWN MEATS

While gene drives may save endangered species from predators and people from parasites, lab-grown meats have the potential to save billions of farm animals and fish from us. Every single day, we slaughter more than two hundred million farm animals and three billion aquatic animals. With cell-based meat technology, all this could change.

Following in the footsteps of Impossible Foods and Beyond Meats, whose plant-based burgers are now featured on the menus of fast-food restaurants, Memphis Meats, a food technology start-up, wants to deliver the real thing, but without harming a single animal. Based in Berkeley, California, this start-up is on a mission to bring meat to the world by manufacturing it in bioreactors.

Memphis Meats isn't alone, there are a number of cell-based meat start-ups, which are working on everything from beef, duck, pork, and chicken to fish. But it's not as easy as dumping a bunch of stem cells into a bucket and waiting for them to reproduce. The meats need to have the flavor, texture, and consistency of real meat. No one wants a mushy steak. That's why scientists have been experimenting with growing muscle cells on a gelatin scaffold.

"Muscle cells need a structure to grow on, the same way the walls of a building need a steel frame, or a house needs a wooden skeleton," says Kevin Kit Parker, a bioengineer at Harvard.[57]

Another problem is that the meat cells quickly replicate at first, but over time they slow down and then stop entirely. To get around this, Memphis Meats is using gene editing to encourage the cells to regenerate.

If this isn't enough on their plate, there's the issue of cost. In 2013 a prototype burger cost more than $300,000. By 2018, Memphis Meats had the price down to $2,400. Their goal is to have the cost at $5 or less per patty.

"The biggest challenge is taking what's in the lab and making it commercially viable," says David Welch, director of science and technology at the Good Food Institute.[58]

With a $1.9 trillion market for meat and lots of venture capital flowing into the sector, whatever problems exist today will be solved. This is good because a typical American consumes more than two hundred pounds of red meat and poultry every year, while China is now the world's largest consumer of meat. The best way to have a sustainable supply of meat without converting a large portion of the population into vegetarians is to ramp up the production of cell-based meats. If it's healthier, cheaper, and just as tasty as farm-raised animals, adoption shouldn't be a problem.

In addition to saving countless animal lives, could lab-based meats save the environment? A single cow produces about 220 pounds of methane per year. Worldwide, there are approximately 1.5 billion cattle. Just do the math. Add in an average of twenty-six gallons of water and twenty-eight pounds of feed per cow every day, and you begin to get an idea of the environmental impact. That said, many lab-grown meats still have a long way to go. Today, the carbon footprint of a typical cell-based chicken is roughly five times that of a farm-raised chicken and ten times higher than plant-based products. This is because the process uses a lot of energy.

Mosa Meat in the Netherlands, however, claims that its cultured meat products produce 96 percent fewer greenhouse gas emissions, use 99 percent less land, and require 96 percent less water than livestock meat. The start-up predicts that once cultured meat becomes a mass-market food, there will be no need for industrial farms.

In a hundred years we may look back at raising and slaughtering animals as a barbarous and inhumane practice, much like we view human sacrifice today. Children may ask their parents, "Why would anyone kill an animal just for food when you can grow meat in factories?"

DNA AND ORGANIC COMPUTERS

Humans are not only figuring out how to grow meat in petri dishes but also how to grow organic computers in test tubes. While traditional computers use silicon chips, organic computers use biological transistors made from genetic materials. These DNA and RNA switches are called transcriptors.

DNA computing was first demonstrated in 1994 by Leonard Adleman. Using only DNA, he solved the traveling salesman problem—that is, how to find the most efficient route for a salesman to take between two locations. Since Adleman's experiment, DNA-based circuits have succeeded in implementing Boolean logic, arithmetical formulas, and neural network computations. Called molecular programming, this field is starting to take off.

The problem is speed. DNA is painfully slow. It can take several hours to compute the square root of a four-digit number. DNA circuits are also single-use and need to be re-created each time they run. With these limitations, will DNA computing ever compete with silicon?

Not right away, but scientists around the world are making progress. MIT and the Singapore University of Technology and Design announced a groundbreaking discovery using an organic virus that could enable the development of faster and more efficient organic computers. Columbia University announced that they'd stored an entire computer operating system on a strand of DNA. Researchers from Microsoft and the University of

Washington have gone on to demonstrate the first fully automated DNA system to store and retrieve data.

"Our ultimate goal is to put a system into production that, to the end user, looks very much like any other cloud storage service," says Karin Strauss, a researcher at Microsoft.[59]

A tiny smear of DNA can hold ten thousand gigabytes of data. This means that a data center as large as a shopping mall could be shrunk down to the size of a sugar cube. DNA is also cheap and easy to synthesize, and computing with DNA requires much less energy than silicon processors. While a Google data center may consume millions of dollars worth of energy in a single year, a biocomputer can operate on just a few cheap metabolites.

In addition to DNA storage, there is potential for organic computers to go places silicon cannot. A team of researchers from ETH Zurich used CRISPR to build functional dual core biocomputers within human cells. Imagine having living computers in our bodies, monitoring our health, repairing damaged tissue, and regulating our bodily functions. We may even have potential to enhance our intelligence with organic computing. DNA computers can also interact with a biochemical environment, enabling the delivery of medicines and therapies inside living organisms.

Israeli researchers have taken the next step, creating biological circuits inside cockroaches. They created DNA folded like origami to make nanobots capable of delivering a payload. The payload might be a molecule, an enzyme, or an antibody. Each payload can activate or inactivate the next nanobot in the chain, thereby creating a circuit inside a living cell. These biocircuits could be used in a number of ways. In one experiment, scientists used biocircuits to identify cells as cancerous or noncancerous. Then they sent the cancerous cells a signal to self-destruct. In the future, biological computing devices similar to this may be used to noninvasively monitor the progression of tumors and target drug treatments to those sites.

This is only the beginning of a long road. As the science matures, researchers will continue to discover new things about how organic computers function and develop more applications uniquely suited to their capabilities.

CHIMERAS, BIOPRINTING, AND TRANSGENIC LIFE-FORMS

It's one thing to modify DNA for use as a computing device and another to combine the DNA of living creatures to make bizarre new hybrid species. But that's exactly what scientists are doing with chimeras.

The word *chimera* has ancient origins. It comes from the Greek name for a mythological beast with a lion's head, goat's body, and serpent's tail. Today, scientists use the term *chimera* to describe any organism containing a mixture of genetically different tissues formed by the fusion of early embryos, mutation, or similar processes.

More than a century ago, author H. G. Wells envisioned this in his sci-fi classic *The Island of Doctor Moreau*. Published in 1896, the story depicts a mad scientist who transforms beasts into humans, sculpting their bodies and brains in his own image. However, the creatures can't escape their brutish nature. Eventually the social fabric breaks down, anarchy erupts across the island, and the chimeras destroy Doctor Moreau and his vision.

Less than a century later, science caught up to fiction. Back in 1980 Janet Rossant published a paper announcing the creation of a chimera that combined genes from two different species of mice. "We showed [that] you really could cross species boundaries," says Rossant.[60]

In 1984 the Institute of Animal Physiology in Cambridge created the first geep by combining goat and sheep DNA. The resulting chimera contained a mosaic of goat and sheep tissue spread throughout its body. For example, the parts of its skin that grew from the sheep embryo were woolly, while those that grew from the goat embryo were hairy.

Now, scientists are looking to create chimeric pigs that can grow human organs. This may seem bizarre and even frightening, but there's a good reason for it. The world is facing a severe shortage of human livers, hearts, lungs, and kidneys. Every day in America, an average of thirty-three people die waiting for an organ transplant. With approximately 115,000 patients stuck on waiting lists in the United States alone, the problem isn't going away.

To address this, Craig Venter, a pioneer of the Human Genome Project, teamed up with United Therapeutics to develop genetically altered pig lungs that can be safely transplanted into humans. Although most people don't realize it, swine and *Homo sapiens* share many genetic similarities. The organs of young pigs are close to the size of human organs. By modifying the porcine genome to make them more compatible, scientists hope to create hybrid organs the human immune system won't reject.

So far, primate studies have shown some initial successes. A baboon managed to survive with a pig kidney for 136 days, and a second baboon with a pig heart lived for nearly two and a half years. And the State Key Laboratory of Stem Cell and Reproductive Biology in Beijing recently announced full-term pig-monkey chimeras.

When it comes to developing transgenic pigs with human cells, scientists are just beginning to understand what's involved. The process typically begins with stem cells that are created from adult human cells. These are introduced into an early-stage pig embryo. The hybrid embryo is returned to the sow, which can then give birth to the chimera.

Another approach is to use gene-editing tools to create host animals that develop without a specific organ. Scientists would then inject human stem cells into the embryo to fill the void of the missing organ. As the embryo matures, a 100-percent human organ would develop inside the chimera.

Despite the optimism of leading geneticists, it has been a bumpy road. Initially, Juan Carlos Izpisua Belmonte, a professor at the Salk Institute, used to believe that using a host embryo to grow organs was fairly straightforward. However, it wound up taking him, along with forty colleagues, nearly four years to figure out how to create a simple transgenic pig, which is only one of many hurdles they now face.

"Even today the best matched organs, except if they come from identical twins, don't last very long because with time the immune system continuously is attacking them," says reproductive biologist Pablo Ross from the University of California, Davis.[61]

Now imagine how hard it is to get the human body to accept an organ from another species, even if it is transgenic. There has to be a large enough percentage of human cells in the animal organs for it to work. However, getting the embryo to accept more human cells is highly problematic. Even if scientists manage this, how long will the organ last before it is rejected or simply fails? There's also the risk of introducing swine viruses.

Egenesis, a venture-funded start-up, is attempting to solve this problem. It's using CRISPR to knock out the virus inside the embryo. Researchers are also experimenting with transgenic sheep, which don't contain harmful viruses.

Scientists are now combining human DNA with all sorts of animals. With each scientific step forward, we're coming closer to *The Island of Doctor Moreau*. "It's not that people are aspiring to create abominations," says Stuart Newman, a cell biologist at New York Medical College. "But things just keep going. There's no natural stopping point."[62]

Given the slippery slope, should governments institute an outright ban on creating all chimeras using human DNA? Which is more unethical: creating transgenic animals and harvesting their organs, or letting people die waiting? Is there even a right answer?

"There's the concern that these types of experiments create morally ambiguous beings," says Insoo Hyun, a bioethicist from Case Western Reserve University. "We know what sheep are, and we know what people are, but what about sheep that have large contributions of human cells or an entire human organ? That's a new thing: Where does it fall on the spectrum?"[63]

In order to sidestep these thorny questions, some scientists are experimenting with growing organs in the lab instead of relying on animals. Human stem cells can theoretically produce any kind of body tissue. They can be transformed into a heart, lung, or liver cells in a Petri dish. However, stimulating stem cells to grow into a fully functional organ outside a living body is another thing. And patients must undergo painful, invasive procedures to harvest the required tissues.

Bioprinting is another approach. It's similar to 3D printing, except it uses biomaterials that combine living cells and growth factors to create tissue-like structures. Researchers at Wake Forest University have successfully implanted bioprinted structures, like muscles and cartilage, into rodents. Meanwhile, Israeli researchers have bioprinted a miniature 3D heart made from human cells.

"This is the first time anyone anywhere has successfully engineered and printed an entire heart replete with cells, blood vessels, ventricles, and chambers," says Tal Dvir, a professor at Tel Aviv University.[64]

This might sound like the answer, but don't expect to order up a bioprinted heart or kidney just yet. There are still major obstacles to overcome. The biomaterials are limited; bioprinted organs have trouble maintaining structural integrity; and the intricacy of natural vasculature is difficult to mimic with the current technology.

So that brings us back to transgenic animals. They may be the best option in the near future for organ transplants, but in addition to the technical hurdles, there are social issues. What happens when scientists begin creating human-monkey chimeras to research cures for brain diseases, like Alzheimer's and Parkinson's? Is the resulting chimera animal or human? When do we cross the line?

What about using the DNA from primates, instead of human DNA? Chimpanzees share roughly 99 percent of our DNA, but they clearly are not human and are regularly used in lab experiments to test new drugs and study diseases. This is a loophole, but is it really more ethical? Scientists are facing these questions right now. In fact, Izpisua Belmonte was offered a $2.5 million grant on the condition that he use primate cells rather than human cells to create his chimeras. He accepted the grant but also continues to work with human cells through other grants.

In a laboratory in China, Izpisua Belmonte and his team developed the first human-monkey hybrid. The researchers created chimeric human-monkey embryos using human stem cells. To address ethical issues, Izpisua Belmonte's research team created mechanisms so that if human cells migrated to the brain, they would self-destruct. They also set the red line at fourteen days of gestation, which means all the human-monkey chimeric embryos were destroyed within the two-week period.

The question remains if other scientists will be so careful. Probably not. Bing Su, a geneticist at Kunming Institute of Zoology in China, added human brain genes to rhesus macaque monkeys, with the goal of improving their short-term memories. Some bioethicists are concerned that chimeras containing human DNA may eventually become a class of subhumans that could be used in lab experiments, sold for their body parts, and exploited in other ways. In a world where our technology is outpacing our ability to implement legal and ethical guidelines, the potential for abuse is enormous.

Under pressure from religious groups, in 2015 the US National Institutes of Health imposed a moratorium on funding for human-animal chimeras. It's now lifting the ban, provided each experiment undergoes a thorough review before funding. But what would this review entail? And who should be empowered to make the ultimate decisions: government officials, scientists, religious leaders, or someone else?

Even if the United States entirely restricts this type of experimentation, what about other countries? Will researchers in China, Russia, Europe, and Japan stop what they're doing? If the United States bans or limits human-animal transgenics, America will probably fall behind because most of the top scientists in the field may feel compelled to relocate to less regulated environments.

There's little doubt among experts that transgenic science is accelerating, and there will be a steady stream of breakthroughs over the coming decades. In the near term, chimeras have the potential to speed up the testing of lifesaving drugs. Human trials are long and expensive. The world's pharmaceutical giants spend billions of dollars on this every year, and many promising drugs are never fully developed because of the prohibitive costs.

"If we were able to put human cells inside a pig's liver, then within the first year of developing the compound, we could see if it was toxic for humans," says Izpisua Belmonte.[65]

This would be a boon for developing the next generation of disease-fighting drugs. Imagine how many people will have their lives transformed if we can speed up drug testing and development. In addition, transgenic animals have the potential to dramatically lower the cost of drug production. Researchers have already spliced human genes into chickens' DNA so that the birds could lay eggs containing an anti-cancer drug. Once the eggs are laid, pharma companies can extract the drug from the eggs, package it up, and put it on the market.

"Production from chickens can cost anywhere from 10 to 100 times less than the factories," says Lissa Herron at University of Edinburgh.[66]

They call these genetically altered animals bioreactors. Scientists are also working on creating bioreactors using milk from transgenic goats, sheep, rabbits, and mice.

Chimeric animals could take many forms in the future, as farmers combine the best traits of cattle, horses, sheep, pigs, goats, camels, and llamas to come up with entirely new types of livestock. NASA may even adopt this technology to develop animals suited for life on other planets.

In China, enterprising scientists are already preparing to sell genetically altered pets. They've developed a new breed of micropigs by applying a gene-editing technology called TALENS. They can produce tiny pigs that weigh less than fifty pounds as adults. Customers will even be able to select the fur color and pattern when ordering. These pigs will sell for around the same price as many other pets.

"We plan to take orders now and see what the scale of the demand is," says Yong Li, a senior director at the Beijing Genomics Institute.[67]

Some people may prefer chimeric pets over traditional breeds. Eventually, it might become possible for pet lovers to design the look and feel of their transgenic cats and dogs. This may include everything from eye color and body shape to the texture of fur and specific personality traits.

Zoos and theme parks might opt to display the latest chimeric creations to draw crowds. Can you imagine what a hippozebra looks like? Theme parks may even produce chimeras to resemble mythical creatures, like a Pegasus, griffin, or unicorn. Who knows?

GENETIC IMAGINEERS: A BRAVE NEW WORLD

What if I told you that you could have muscles like Arnold Schwarzenegger, change the color of your skin, or cure an ailment, and all you had to do was inject yourself with DNA? And what if you could do this at home with a gene-editing kit that you could buy online? Believe it or not, biohackers are attempting this.

Josiah Zayner, a Silicon Valley biohacker and social activist, is one of those rebels. With his constantly changing hair color and antiestablishment views, he looks the part of someone willing to push the boundaries of science and the law. After graduating with a PhD in biophysics from the University of Chicago, Zayner went to work for NASA Ames Space

Synthetic Biology Research Center, helping to design habitats for future colonies on Mars. This may sound like a dream job, but Zayner wasn't satisfied. He quickly became "fed up with the system" and the slow pace of researchers just "sitting on their asses."[68] So, after only two years, he quit his NASA fellowship to do his own thing.

Fascinated by the potential of gene editing, he launched a crowdfunding campaign to sell inexpensive CRISPR kits to anyone who wanted to experiment with editing bacterial DNA. Working out of his garage, he also figured out how to inject frogs with CRISPR DNA that enhanced their muscles. When it appeared to be working, he began selling these kits online to anyone curious enough to give it a try. It would have been fine if he'd stopped there, but Zayner isn't that type. Perpetually restless and always wanting to challenge authority, he decided to perform a publicity stunt. While livestreaming at a biotech conference, he injected himself with a CRISPR DNA cocktail designed to knock out the muscle-suppressing protein myostatin.

The stunt worked. The media went crazy, and he became an overnight celebrity in the world of biotech. At the same time, it backfired on him, as many of the leading scientists in the field condemned his reckless behavior and state officials took notice and began investigating his actions. He has since gotten entangled in numerous disputes with the government and the scientific community over his actions.

Zayner now admits that he went too far: "There's no doubt in my mind that somebody is going to end up hurt eventually. Everybody is trying to one-up each other more and more. It's just getting more and more dangerous."[69]

Not only did Zayner set a bad precedent, but his experiment didn't seem to yield the intended results. There's no evidence it actually worked to enhance his muscles, and it may have caused lasting harm to his health. Time will tell. But that hasn't stopped Zayner or those like him.

Biohacker activist Gabriel Licina, from the group Science for the Masses, believes that all Americans should have the right to experiment on themselves. To prove his point, he decided to give himself night vision. Taking a chemical that allows deep-sea fish to see in the dark, he squirted it into his eyes. If this sounds risky, it is. No one knows for certain what are the long-term consequences. Scientists point out that increased light amplification may have adverse effects on the cellular structure of the eye.

After taking the chemical, Licina claimed he could see more than 160 feet in the dark for a brief time. If you trust Licina's data, it's an interesting experiment, but he could have done the same test on a lab animal at much lower risk to himself and with much less controversy. That said, what activists like Licina and Zayner want most is publicity for their cause. They feel that too much power is in the hands of the government and big corporations, and they're on a mission to bring biotech directly to the people.

One of the examples they give of what's wrong with the current system is how corporations often withhold valuable treatments from the general public by overcharging for them. This means only the wealthy or those with very good insurance can afford the treatments, leaving others to suffer unnecessarily.

An example of this is the case of Sofia Priebe, a fourteen-year-old who was slowly going blind because of a genetic mutation that caused her retinas to deteriorate. That's when her parents heard of Spark Therapeutics. The company has used CRISPR to develop a gene therapy that can restore eyesight. It's like a miracle for those affected, but it comes at a high price: $425,000 per eye. Unfortunately, Sofia's parents couldn't afford this, and they didn't have the insurance to cover it. This meant they were faced with the prospect of having their daughter go blind.

When asked why the treatment was priced so high, Jeff Marrazzo, CEO of Sparks Therapeutics, says, "It came down to the value we believed was inherent in the therapy."[70]

By "value," Marrazzo means the amount they could extract from the market. It's what people and insurance companies are willing to pay. It's similar to profiteering in wartime because the buyers have no choice. What parents would let their daughter go blind, even if it meant losing their home and their life savings? Ironically, much of the CRISPR research, on which this treatment was based, originally came from government-funded projects.

After a lot of negative press, Marrazzo said, "We are committed to finding a novel solution to providing an installment payment option."[71]

Does this mean some people will become indebted to biotech companies for huge sums of money they may never be able to repay? And what about people on Medicare or Medicaid? Should the government pay full price or deny patients treatment? Spark Therapeutics isn't alone in charging high prices. The entire drug industry has moved in this direction, as it works to maximize profits at the public's expense.

This is why biohackers like Zayner believe it's their duty to take the technology directly to the people. Aaron Traywick launched Ascendance Biomedical with this in mind. Like Zayner, he believed in flaunting FDA regulations and going it alone. Traywick made headlines when he injected himself with a DIY herpes treatment in front of an audience at a self-experimentation conference. His start-up had previously livestreamed another self-injection. Tristan Roberts, a twenty-eight-year-old computer programmer who is HIV positive, received a compound provided by Ascendance that was supposed to lower the number of HIV particles in his blood.

"I want to dedicate this to all the people who have died while not being able to access treatment," said Roberts, before injecting himself with this untested drug. A few weeks later, he was worried that he'd made a big mistake. "I'm 98 percent certain this is fine. But there's still this 2 percent that's like—this could be something terrible."[72]

In the end, nothing much happened. Roberts didn't die, and the treatment didn't cure his HIV. Later, he wound up having a falling-out with Traywick and the treatments stopped. Traywick, with no formal medical training and very little in the way of funding, moved on to another experiment. His next idea was to use CRISPR to treat lung cancer patients at a clinic in Tijuana, Mexico.

At this point, we have to ask the question: Is this type of rapid, low-budget, unregulated experimentation a good thing or a bad thing? Does it help push science further or unnecessarily put people's lives and health at risk?

"This is the kind of thing we were afraid would happen, and it is probably destined to happen," said Michele Calos, vice president of the American Society for Gene and Cell Therapy and a researcher at Stanford University.[73]

In the end, the experiment in Tijuana was canceled after Traywick died at age twenty-eight in a sensory deprivation tank in a spa in Washington, DC.

Zayner continues to conduct experiments. However, the California Medical Board has investigated him for practicing medicine without a license. The FDA has gone after Zayner, objecting to him selling gene-editing kits for use on humans. This is against the law.

Zayner tweeted, "The fucked-up part is that so many people are dying not because of me but because the FDA and government refuses to allow people access to cutting edge treatments or in some cases even basic healthcare. Yet I am the one threatened with jail."[74]

Regardless of what you think about Zayner and Traywick, CRISPR is here to stay, and it's readily accessible to anyone with the knowledge and few hundred bucks to buy a gene-editing kit. The number of regulated and unregulated CRISPR trials around the world is rising. Human trials are underway on everything from curing kidney cancer to eliminating sickle-cell anemia.

In China, He Jiankui, a researcher at the Southern University of Science and Technology in Shenzhen, used CRISPR to edit the embryos of twin girls in order to make them HIV resistant. When the world found out, his actions were condemned. The Chinese government ordered the suspension of all He's research projects, and the university fired him. Later, Shenzhen courts sentenced He to three years behind bars and a ¥3 million ($430,000) fine.

The reason for the outcry and harsh punishment is that this type of experiment could lead to lasting genetic disorders for the girls and their future offspring. CRISPR is an imperfect tool. Trying to edit one gene can create unintended alterations elsewhere in the genome. The babies were born healthy, but He's team has admitted to finding one "off-target" mutation so far.

He may have pushed the limits of HIV research, but at what price? It's one thing to experiment on yourself, like Zayner did. It's another thing to provide untested drugs to consenting adults, like Traywick did. But should anyone be allowed to experiment on unborn children who have no say in the matter? He literally put his own research and ambitions ahead of the children's safety, which breaks the Hippocratic Oath. The question is, What should governments do about this? Should all experimentation with human embryos be outlawed?

What if I told you that you could pick the sex and eye color of your next baby? Would you do it? Well, you can do it right now. It's legal in the United States. Jeffrey Steinberg, director of the Fertility Institutes in New

York, lets parents choose not only the sex of their unborn children, but also whether their eyes are blue, brown, or green, assuming the trait is present in their genes.

"The technology was out there. It was being applied only to diseases," explains Steinberg. "I've decided to open the door and expand it and say, 'Listen, this is something that people are interested in, causes no harm, makes people happy. Let's expand it.'"[75]

Using preimplantation genetic screening, developed more than two decades ago and an offshoot of in vitro fertilization, couples can now dive deeper into their baby's genetic makeup and choose which of their eggs they prefer to have implanted. The embryos aren't being modified. Parents are just being given a simple choice: Do you want the egg with green eyes or brown eyes?

"People call up asking for all kinds of things: Vocal ability, athletic ability; height is a big one. I have a lot of patients who want tall children," says Steinberg.[76] "Women from countries like China and India tend to want boys. In the rest of the world, there's a slight preference for girls."[77]

Where do we draw the line? Look at the problems in China, where they now have thirty-four million more men than women. India has a similar issue. Social pressures are so strong that even strict government regulations haven't been able to prevent the imbalance. And there are the issues around eugenics. Should parents be able to choose an embryo with DNA that will produce a lighter skin and hair color? What does this mean for society?

"If you do what I do, you can't have a strong ethical opinion," admits Steinberg, unless, of course, the parents ask for "something that is going to be harmful."[78] Despite this stance, his clinic stopped offering pigmentation as an option after public outrage became too much to handle.

Even if the US government bans fertility clinics from engaging in this practice, other governments most certainly won't. John Zhang, a medical scientist, made headlines in 2016 by successfully producing the world's first three-parent baby. His goal was to enable parents to conceive a child without passing on metabolic diseases caused by faulty maternal mitochondrial DNA. A US National Academy of Sciences study concluded it could be ethical to attempt the procedure for this purpose. However, because the

FDA won't permit it in the United States, Zhang had to move his practice outside the country. What's clear is that couples who have the resources will go wherever it is legal to get these more controversial procedures done. Right now, if you want to have a baby with three parents, you can go to the Ukraine and get it done. The cost is around $15,000.

"What we're seeing is a fast slide down a very slippery slope toward designer babies," says Marcy Darnovsky, who heads the Center for Genetics and Society, a US-based watchdog group. "We could see parents feeling eager to give their children traits like greater strength, needs less sleep. Some people are saying that 'Yes, there are genes for IQ and we could have smarter babies.'"[79]

For many, this would be considered a good thing. Who wouldn't want a baby that will never suffer from cancer and other diseases? Why not ensure that your child isn't born autistic or with some deformity? How about boosting your kid's IQ by ten points? Why should we expect parents to accept the genetic lottery when we have the power to determine the outcome?

The reason is because social pressures may force parents to conform. Nobody wants to have the least intelligent child in the class. But what if certain families can't afford these treatments, while others can? The problem is that even if this technology is outlawed in one country, parents with money may travel overseas for the procedure. This could lead to a troubling situation where certain segments of the population get left behind.

Do we want a world where the wealthy are able to endow their children with the highest IQs, good looks, perfect health, and other traits that allow them to outcompete everyone else? In essence, we could be entering a world where the rich are able to create a master race that winds up controlling the rest of us.

Being Jewish and having lost distant relatives in the Holocaust, I've thought a lot about this. It's very disturbing, but there's no way to turn back the clock. The technology is here now, and it's only going to get better. As a society, we need to begin to grapple with these issues and formulate a strategy, so this type of scenario doesn't fully play itself out.

The first thing we need to recognize is that entirely banning designer babies is not possible. All it takes is for a single country to see it in their national interest to have smarter, healthier, and more capable babies. As soon as a country makes it legal, not only will its citizens benefit, but it will lead the way in creating a new multibillion-dollar industry.

This type of scenario has already begun playing itself out. Even though the European Union, United States, and China have banned creating CRISPR babies, a scientist in Russia claims he's moving forward. Denis Rebrikov, a researcher at the Pirogov Russian National Research Medical University, said he plans to implant gene-edited embryos into women.

"This is irresponsible," says R. Alta Charo, a bioethicist at the University of Wisconsin, Madison. "My biggest worry is that he's going to bring about the birth of children who are going to suffer because he wanted to play around."[80]

"The technology is not ready," agrees Dieter Egli, a genetic researcher at Columbia University.[81]

Even if the technology isn't ready now, eventually it will be. At that point, the world will have a decision to make. Let's assume that we can get every single country in the world to sign a pact prohibiting designer babies, which is unlikely considering our track record on nuclear arms control and climate change. Will this be enough to keep doctors like Rebrikov from offering their services? I don't think so. More likely, a black market for designer babies will emerge. Where there's a demand and a lot of money, someone is always willing to provide the service. No matter how it plays out, it's hard to imagine a scenario where the wealthy won't take advantage of the opportunity to have their children enhanced.

The real question is not whether the world will create a new superspecies, but when and how. We have to recognize that fact and make sure it happens in a way that is equitable and good for society. That will probably mean legalizing and regulating it, much in the same way we do countless other medical procedures. In fact, this will be the only way to drive down the price so that it becomes affordable for all segments of society. Governments may even wind up having to provide subsidized gene-editing services for the poor.

It's not hard to imagine this becoming a campaign slogan in the future. Along with universal basic income and universal health care, there will be universal gene enhancement. If done right, we may find ourselves in a world where everyone receives the latest gene upgrades before birth, as well as throughout their lifetimes. Having your genes updated could become as routine as getting a flu shot.

Even with broad access, many of the ethical issues won't go away. There will still be questions of whether the wealthy can afford superior upgrades to what the general populace receives. Also, what types of upgrades should be permissible? Should parents be allowed to go so far as to design their children's personalities? It's one thing to have a smarter child, but what about creating a more docile or competitive personality?

What will these changes mean for the human species? Will taking natural selection into our own hands diminish diversity? Being depressive or bipolar isn't something most people want their children to experience, but some of our greatest artists and thinkers have suffered these conditions, including Ludwig von Beethoven, Leo Tolstoy, Charles Dickens, Vincent van Gogh, Winston Churchill, and Mariah Carey. If we begin to systematically eliminate these variations from the population, will we be curbing our own potential for creativity and innovation in the future?

What happens when governments decide it's in the best interests of society to promote certain characteristics in their citizens? Certain countries may decide to engineer their citizens to be more compliant and law-abiding. Others may go further. Could government turn humans into cookie-cutter replicas of the model citizen?

We may wind up limiting the human experience to just a narrow subset of socially acceptable personality types. It may be annoying to interact with an outspoken contrarian at work, but do we want a world where no one speaks up?

What is a perfect person? And do we want everyone to become that? There's a beauty to the randomness of the genetic lottery. It's what makes the world so interesting.

FORCE 3
HUMAN EXPANSIONISM

Human Expansionism, the force driving human beings to push to the edges of our known universe, will propel us further into the quantum world and deeper into outer space, in order to harness their vast potential.

Is it the destiny of humankind to become masters of the universe? That remains to be seen, but we have become masters of our planet. We have conquered all competing animals, and now we are pushing further into unknown territory. We are drilling down into the subatomic world of quantum physics, while at the same time extending our reach into outer space. As long as the human imagination can conceptualize it, we will go there. We can't help but push the boundaries of the known world out further, and what we've found so far is that there is always more beyond what we previously thought existed.

Let's begin by entering the subatomic universe of nanotechnology and quantum physics. We'll examine how this invisible realm has the potential to unleash massive changes in everything from quantum computing to producing biofuels, curing cancers, and constructing skyscrapers. Are miracle materials, like graphene, poised to spawn entirely new industries? Can scientists create sophisticated robots out of nanoparticles, and what will these molecular machines be capable of doing? Are there any risks in manipulating subatomic particles? And what happens if this powerful technology falls into the wrong hands?

Next, we'll rocket into space and look at the feasibility of colonizing the Moon and Mars. We'll discuss the viability of far-out ideas, like using warp drives, space elevators, hibernation chambers, and quantum teleportation to conquer the vastness of space. We'll also bring up the ethical concerns, like should we allow private companies to place billboards into orbit? Is leftover space junk endangering our astronauts? And is it possible for colonists to avoid contaminating planets like Mars with invasive bacteria?

These are just some of the topics we'll consider, as we journey to the limits of human expansionism.

QUANTUM COMPUTING

The story of quantum computing began back in the early 1980s, when Paul Benioff, a theoretical physicist at Argonne National Laboratory, proposed a quantum mechanical model of the Turing machine. Later, Richard Feynman and Yuri Manin postulated that quantum computers could simulate things that classical computers could not. Since then, scientists have figured out how to use the quantum-mechanical phenomena of superposition and entanglement to perform computations.

Here's a simplified explanation of how it works. Qubits are analogous to bits in traditional computing, except that they aren't limited to binary states. A qubit can be in a 1 or 0 quantum state—or a superposition of the 1 and 0 states. The strange part is that whenever a scientist measures a qubit, it is always either a 1 or 0. However, the probabilities of being a 1 or 0 depend on the quantum state that the qubit was in before measurement. In this way, they are completely different from silicon-based transistors, and this enables them to perform certain calculations at exponentially faster speeds.

Taking advantage of this quantum phenomenon, IBM, Google, and others are now attempting to build the most powerful computers the world has ever seen. Here's an example of how beneficial quantum computing can be: when IBM developed the Deep Blue AI that defeated chess champion Garry Kasparov in 1997, it was able to examine two hundred million possible chess moves every second, giving it a significant advantage. In comparison, a typical quantum-based system should be capable of calculating around one trillion moves per second.[1]

Using its quantum computer, Google was able to prove the randomness of numbers produced by a random number generator. This extremely difficult calculation would have taken the world's fastest traditional supercomputer 10,000 years to complete. It only took the quantum computer 3 minutes and 20 seconds. This means the quantum system is more than 1.5 billion times faster than a classical supercomputer when performing this particular task. Google's experiment was a milestone because it demonstrated quantum supremacy, which means that a quantum computer can solve a problem that a conventional computer is unable to do in a realistic human time frame. IBM later disputed this, but then a Chinese research group's quantum computer solved an even more difficult calculation in record time. So, one way or another, we're on the road to quantum supremacy.

Another area where quantum computers excel is at processing unstructured data. Computers, corporations, governments, and every person on the planet that's connected to the internet are generating new data at an astonishing rate. More than 2.5 quintillion bytes of data are created every day, and that number is growing as we spend more time online and connect more devices to the internet. Processing and making sense of all this data is a gargantuan undertaking, and that's where the speed of quantum computers comes into play. With quantum computers, we may soon be able to extract actionable information from unstructured data that is currently just sitting idle in databases. This could have an enormous impact on the area of artificial intelligence, which uses data as its raw material, leading to a dramatic advance in AI products and services.

Despite all the progress being made, you won't find a quantum computer on your desktop right away. Not only is the current generation of quantum computers bulky, but they are incredibly fragile. Even small vibrations or electromagnetic waves in the surrounding environment can impact the superconducting metals that entangle the processor's qubits and cause decoherence—that is, when a quantum system loses its state before it can be measured. For this reason, quantum computers need to be kept at extremely cold temperatures. This is why the inside of the D-Wave quantum computer is negative 460 degrees Fahrenheit, close to absolute zero. No one is going to be walking around with a quantum smartphone anytime soon.

Quantum computers also aren't suitable for the type of work most people do on their phones and laptops. They operate on a completely different set of principles and are better suited to solving extremely hard mathematical problems. This includes things like finding very large prime numbers or simulating how molecules behave. Running photo sharing, messaging, and productivity apps is not an ideal task for a quantum computer. In fact, the apps that most of us use on a daily basis are much better suited for traditional microprocessors.

For the foreseeable future, quantum computers will remain in datacenters and operate in the background. People may not even be aware when they are tapping into a cloud-based service powered by a quantum computer. But these computers will have a profound impact on the entire internet and everything we do. Let me give you a few examples.

First, quantum computers will soon make our current encryption obsolete. They will be able to crack any conventionally encrypted message in seconds. This could render our entire financial system vulnerable, from our banks to the stock market, as well as corporate and government databases. The first ones to produce quantum computers will literally have the keys to the kingdom.

It's scary to think of what may happen if this technology falls into the wrong hands. That said, there's a bright side. Quantum computers will also have the ability to make our information far more secure, by enabling quantum-based encryption. This is why companies and governments around the world are in a race to build the first quantum computers. Using quantum entanglement, there's the potential to create virtually unbreakable quantum networks, which could lead to a quantum internet that is far safer and faster than today's internet.

Quantum computing may have the biggest impact when it comes to science. It could help us unravel many of the mysteries that have eluded researchers because the problems are beyond the reach of conventional computers. In the years to come, we'll see quantum computers excel at modeling molecular interactions, particle physics, astrophysical dynamics, and genetic mutation patterns. This will yield breakthroughs in everything from developing new types of gene therapies and pharmaceuticals to more efficient fuels and solar cells.

Keep in mind that we are only at the beginning of the age of quantum computing. The computers of tomorrow will run circles around what we have in the labs now. Moore's law says that the processing power of conventional computers will double roughly every eighteen months. This has held true for nearly half a century. Will the same be true for quantum computers? If Hartmut Neven's law is correct, quantum computing power will far exceed this. Neven expects a "doubly exponential" rate. If you look at a graph, the growth rate appears to be nearly vertical, which is mind-boggling.

Eventually, we'll see quantum computers completely alter the way our civilization functions. They will be able to model economic market forces, predict weather patterns, and aid in the development of medicines, nanotechnologies, and new materials we could only dream of. The possibilities are so numerous that we may need a quantum computer just to figure out what quantum computers can do for us.

THE BIRTH OF NEW MATERIALS

For millennia, humans have relied on materials we could find and harvest from the natural world. Over time, we have learned to create new materials that never existed before and use them to construct our modern civilization. Now, with tools like AI, faster processors, and quantum computing, we are on the cusp of unleashing a wave of new materials that can have properties that appear almost magical.

Take, for example, graphene, a single layer of carbon atoms arranged in a two-dimensional hexagonal lattice. This 2D material is unbelievably thin and light. It's one million times thinner than a human hair. A sheet of graphene large enough to cover a football field would come in under one gram, less than half the weight of a penny. And it's so flexible that in the wind, it will blow about like a piece of cloth. Despite this, graphene is two hundred times stronger than steel. A single sheet of graphene could theoretically support the weight of an elephant without breaking.

Why is graphene so strong? It comes down to quantum physics. When arranged in a honeycomb lattice, each carbon atom shares a covalent bond with three of its neighbors. This means the atoms share electrons, increasing the material's strength exponentially.

Another property of graphene is that it's the most efficient conductive material known, besides a superconductor. It's better than silver, copper, or any other metal. It's such an efficient conductor that a very small volume of graphene can store a tremendous amount of energy.

The problem is that it's difficult to produce a large enough sheet of graphene to do many of the things industry would like. Graphene is strong but extremely brittle. A tiny crack in a piece of graphene is enough to weaken it, and cracks form easily during the manufacturing process. This is why we don't see graphene being used to construct bridges or office buildings.

Despite this, graphene is already being incorporated into a growing number of products. Tiny pieces of graphene are being mixed in with other materials, like rubber, to make wear-resistant automobile tires. Graphene is also being blended with cement, enabling it to conduct electricity. Not only could this replace the need for electrical cables inside walls, but since graphene dissipates heat, it could be useful for structures that need to stay cool.

Scientists are now developing textile fibers that combine graphene with various polymers. These graphene fibers might be used to make clothing warm up or cool down depending on skin temperature and other factors. Graphene clothing may also be used to recharge cell phone batteries, monitor heart rates and blood pressure, and even perform medical diagnostics. Other possible applications may include repelling bacteria and mosquitoes, detecting toxic gases in the environment, and reducing body odor. If this isn't enough for you, graphene could trigger color changes in the fabric that sync with your breathing, body movements, or other input.

In one experiment, researchers in Italy added graphene and carbon nanotubes to spiders' drinking water to see how it would affect them. The arachnids wound up spinning webs of graphene spider silk, creating one of the strongest materials on Earth. It's even stronger and lighter than Kevlar, which is used in many bulletproof vests. Imagine having a shirt made with graphene spider silk that could stop bullets, or using graphene silk to repair extensive nerve injuries. There is even the possibility of applying the same technique to other animals and plants, leading to a whole new class of bionic materials. Are you up for indestructible thermal underwear made of graphene-enhanced cotton or wool? How about long-lasting graphene-leather shoes or scratch-proof graphene-wood floors?

Scientists have even discovered that sheets of graphene can produce energy. Researchers observed that graphene sheets were rippling due to ambient heat at room temperature. By harnessing this energy, they believe they can produce a nanoscale power generator. This could potentially be inserted into any device from a smartphone to a fitness tracker to generate a constant stream of electricity without ever needing to be recharged. There's even the possibility of creating self-powering bio-implants, including pacemakers and cochlear implants, that would run off body heat.

Graphene isn't the only game in town. There are other new materials coming that could make it look like yesterday's wunderkind. These include borophene, which is stronger and more flexible than graphene; transparent alumina for skyscrapers; and shape-shifting liquid metal, which may be used to produce electronic blood. There's even metallic foam that could be employed to build floating cities.

This is a small slice of what's emerging from research laboratories around the globe. Scientists are constantly generating new materials and experimenting with them. Eventually, the idea of wearing clothing made of plants or animals, building cars out of steel, and constructing houses out of cement, brick, and wood may seem prehistoric. As we venture deep into the quantum universe of molecules and nanoparticles, humans will continue to expand the boundaries of the material world and reimagine what's possible.

NANOSCALE: MANIPULATING MOLECULES AND ATOMS

To understand nanotechnology, you first need to grasp what it's like to operate at nanoscale. Nanoscale is a thousand times smaller than the microscopic scale and a billion times smaller than the world of meters that we experience every day. Scientists developing nanotechnologies often individually manipulate atoms and molecules. But doing so is difficult. Imagine trying to build a tiny machine or structure one atom or molecule at a time. That's why they've turned to self-assembling nanoparticles, where the components of a system assemble themselves to form larger functional units. This allows them to build much more complex nanostructures.

Another approach to building nanosized objects is to shrink down larger objects to nanoscale. It may sound like something out of the movie *Ant-Man*, but at MIT researchers have concocted a system that can produce 3D nanoscale versions of much larger objects, reducing them to around a thousandth the size of the originals. The objects are covered in a special type of hydrogel; then, using implosion fabrication, the whole structure is shrunk down.

"People have been trying to invent better equipment to make smaller nanomaterials for years, but we realized that if you just use existing systems and embed your materials in this gel, you can shrink them down to the nanoscale, without distorting the patterns," says Samuel Rodriques, a coauthor of the study at MIT.[2]

Scientists are also creating nanostructures using 3D molecular printers. With computer-aided design software, a molecular printer can assemble functional molecules piece by piece. This might be used to create an army of nanobots or assemble a new type of nanomaterial. It may even be used to create precision medicines designed around an individual's DNA. Someday, you may walk into a pharmacy where your doctor has uploaded your DNA and walk out a few minutes later with a drug tailored just for you.

Most people don't realize it, but nanotechnology is already producing our everyday products. Many sunscreens now include nanoparticles. These absorb dangerous ultraviolet light. They also spread more smoothly over the skin. Similar nanoparticles are being used in food packaging to reduce UV exposure and prolong shelf life. And some plastic bottles used for carbonated drinks now contain nanoclays, which helps to increase their shelf life by several months.

Nanotechnology is also being widely deployed in airports, laboratories, and industrial sites to detect chemicals at extremely low levels. A nanosensor can identify a single molecule out of billions. This is useful for detecting illegal drugs at airports, substances in the body, or leaks at a chemical plant.

You may have seen the YouTube videos of the nanoscale fabrics that can repel water, coffee, ink, oil, and even sweat. Nanolayers of positively and negatively charged films can repulse everything from bacteria to acids. This is being used to produce rugs, clothing, and furniture that won't stain. Also, this technology can be used in hospitals to prevent bacterial infections and in hazmat suits to keep workers safe from toxic chemicals.

If nanoparticles can repel bacteria and viruses, how about a nanocondom? Using a nanofabrication technique called electro-spinning, scientists have produced a fabric woven from sperm-blocking fibers knitted together with anti-HIV drug delivery fibers. The result is the perfect female condom. It prevents pregnancy, guards against HIV transmission, and dissolves within hours of use. The Bill and Melinda Gates Foundation has awarded the research team $1 million to bring the product to market.

Another innovation is self-healing nanoplastics and nanometals that bleed when cut. When damage occurs, nanocapsules inside the material rupture and fluid seeps out, repairing the wound. This means you may not need to worry about your car getting scratched because the paint will heal itself. This could also be useful for building anything from electronics to spacecraft. For example, if a chip implanted inside the body becomes damaged or corroded, it could heal itself, avoiding surgical removal. The same is true if we send probes to distant planets, like Mars; the parts and electronics can self-repair.

On the horizon is another breakthrough. Researchers are now developing nanochips, as opposed to microchips. These tiny chips could be a hundred times more energy efficient and won't require any extra electrical power to store memory. This combined with their small size could create an entirely new class of much smaller, lightweight electronic sensors and consumer electronics. They'll also be faster. This is because nanotech could enable transistors to send electrical currents through narrow gaps of air, like a vacuum, instead of silicon.

"Imagine walking on a densely crowded street in an effort to get from point A to B. The crowd slows your progress and drains your energy," says Sharath Sriram, a professor of nanotechnology. "Traveling in a vacuum on the other hand is like an empty highway where you can drive faster with higher energy efficiency."[3]

Sriram didn't come up with this breakthrough on his own. He works with his wife, Madhu Bhaskaran. Both are professors at RMIT in Melbourne. They met at college in India, went to Australia at the same time, obtained master's degrees on the same day from the same department, under the same professor, and then went on to get their PhDs on the same day. Now they run the university's lab together—except, of course, when Bhaskaran was out on maternity leave, which was a real challenge for a couple that's been operating as a single unit for years.

Improving processor speeds is one thing, but what about using nanotech to address some of the world's most serious problems? With global warming, obtaining drinkable water is becoming more of an issue, especially if you live in a dry climate. To mitigate this, researchers have figured out a way to pull water from the air using nanofiber cloth. Mesh nets are able to capture between 2 and 10 percent of moisture in the air, depending on the material used. Theoretically, these means could gather nearly fifty gallons of water per day for every square meter of material.

"This work is to address a humanitarian crisis," says Shing-Chung Wong, a professor at Ohio's University of Akron. "The ultimate goal is to provide a viable solution to help those areas around the world which are affected by drought. In my opinion, every human being is entitled to fresh water; not just the richest people globally."[4]

Another problem the world faces is greenhouse gases. Ajayan Vinu is a professor of materials science and nanotechnology at the University of Newcastle in Australia. He was born in a small village in Tamil Nadu, India, and has become one of the top fifteen leading nanomaterial scientists in the world. Although it seems unbelievable, he has figured out a way to capture CO_2 and convert it into fuel. He says it can clean the environment and generate energy at the same time. Vinu's system uses nanoporous carbon nitrate to capture CO_2 and then converts it to fuel with the help of sunlight and water. His team is now working to couple this with solar cells and battery technology. "It is a device that can run any vehicle and provide energy without the need for fossil fuels," says Vinu.[5]

Nanotechnology is getting to the point where it can perform what seems like magic. If you're a Harry Potter fan, you'll appreciate this—nanotech researchers have developed an invisibility cloak that is only eighty nanometers thick. It uses blocks of gold nanoantennas to reroute reflected light waves so that an object appears invisible. Although this cloak is only microscopic in size, the principles behind the technology should enable it to be scaled up. "This is the first time a 3D object of arbitrary shape has been cloaked from visible light," says Xiang Zhang, a director at Berkeley Lab.[6]

If you don't like the idea of people spying on you with invisibility cloaks, there's a nanosolution to that, too. Scientists have succeeded in equipping mice with infrared vision by injecting tiny nanoparticles into their eyeballs. This essentially gives the mice super vision, without seeming to interfere with their normal sight. "In our study, we have shown that both rods and cones bind these nanoparticles and were activated by the near infrared light," says Tian Xue of the University of Science and Technology of China. "So, we believe this technology will also work in human eyes, not only for generating supervision but also for therapeutic solutions in human red color vision deficits."[7]

The possibilities for nanotech are endless. The more scientists learn, the more applications will emerge. It won't be long before nanotech endows us with wondrous powers. Eventually, we'll be able to command legions of unseen nanosized robots to march around our homes, sweeping up dust and devouring crumbs; trek across our skin and clothes, removing odor-causing bacteria and dirt; roam our city streets, patching cracks and fixing potholes; swim through our sewage systems, unclogging and repairing pipes; crawl over our cars and buildings, scrubbing clean windows and walls; and even navigate through our bodies, fighting off disease and repairing cells. These tiny robot soldiers are the vanguard of human expansionism, as we master the invisible world of molecules.

NANOBOTS: MINIATURE MEDICAL MACHINES

Can you imagine walking around with thousands, or even millions, of tiny nanobots in your blood, organs, and muscle tissue? If the thought makes you uncomfortable, you should know that only 43 percent of the cells in your body are human cells. The remainder are microscopic colonists. These include trillions of bacteria, viruses, fungi, and archaea. So, adding a few million nanobots to the mix might not be such a big deal, especially if it can improve our lives and health.

Nanotechnology is already in the process of remaking health care in myriad ways, from new types of drug delivery mechanisms to nanotools used in hospital operating rooms. Researchers have even developed nanoparticles that when injected into your bloodstream will begin to eat away your insides. No, this isn't a Michael Crichton novel; it's actually a good thing. Developed at Michigan State and Stanford, these nanoparticles target harmful plaques that lead to heart attacks, strokes, and other fatal diseases. By removing plaques, the nanoparticles can clear your arteries. With this, you can chow down on as many double cheeseburgers as you like without feeling guilty.

If you wind up in the operating room in need of a triple bypass because you failed to inject your artery-cleaning nanobots, how would you like to have an electric bandage? Researchers have developed a bandage that uses nanogenerators to apply electrical pulses to a wound to help accelerate healing. The method leverages energy generated from a patient's own body motions, and the cost isn't much more than a regular bandage. This technology is especially useful for serious wounds and internal bleeding.

"We were surprised to see such a fast recovery rate," says Xudong Wang, a professor at the University of Wisconsin. "We suspected that the devices would produce some effect, but the magnitude was much more than we expected."[8]

Even electric bandages aren't always enough to stop bleeding. The blood also needs to clot. There are times when blood won't clot properly, like with serious wounds or patients with hemophilia. Natural platelet products from human donors are in limited supply, have short shelf lives, and have potentially severe biologic side effects. To solve these problems, researchers developed synthetic nanoparticles that can mimic platelets' abilities to clot at the site of a bleeding injury. When injected into the body, these synthetic platelets can treat bleeding complications from surgery, trauma, and low platelet counts.

"Our nanoparticle technology can be used in civilian and military scenarios of traumatic non-compressible bleeding where donor platelets are not readily available," says Anirban Sen Gupta, a professor at Case Western Reserve.[9]

Now for my favorite nanomachine of all. Are you up for a mouthful of robots? Researchers at the University of Pennsylvania are using microrobots armed with nanoparticles to crawl across teeth and remove dental plaque. Children in America may soon rejoice at never being nagged again about brushing and flossing. If it works, you may even be able to skip going to the dentist altogether.

If never having to brush your teeth again sounds too good to be true, what about getting rid of cancer? Researchers at KU Leuven in Belgium discovered that tumors are sensitive to copper oxide nanoparticles. They also found that nanoparticles containing iron oxide could target cancer cells without touching healthy ones. They put these together to develop a new immunotherapy, which they believe may replace chemotherapy in the battle against lung, breast, ovarian, and colon cancers.

When developing this treatment, the researchers ran experiments on mice and noticed that copper compounds could not only kill the tumor cells directly but also assist those cells in the immune system that fight foreign substances. The combination of the nanoparticles and immunotherapy made the cancer tumors disappear entirely in the mice. The next step is to test other metal oxides and find out which ones most efficiently fight off cancer cells with long-lasting immune effects. If this works, the researchers believe this technique could be applied to 60 percent of all cancers.

At Arizona State University scientists have taken a different approach. Hao Yan, a specialist in the field of DNA origami, designed a fleet of nanobots to seek out and destroy cancerous tumors, while leaving healthy cells unharmed. These nanobots act by targeting the tumor's blood supply and blocking the flow. Since all tumors require blood to survive, this technology has potential to combat a wide variety of cancers.

Yan likes to say that if he wasn't a scientist, he'd be a rock star, playing electric guitar and singing on stage. He puts his creative abilities to use in his lab, attracting top talent and fostering an environment where they can thrive. "I throw them in the pond and let them swim," says Yan. "I don't want to produce a technician. I want to produce a creative thinker and a scientist who can come up with their own ideas and solve problems on their own."[10]

Three researchers in his small lab have already been named "New Innovator" by the National Institutes of Health, and in 2019 Yan was selected as one of *Fast Company*'s "Most Creative People in Business." Yan and his team appreciate that, working together in an interdisciplinary fashion, they are able to combine nanotechnology with fields like tumor biology and cancer immunotherapy: "I think we are much closer to real, practical medical applications of the technology. Combinations of different rationally designed nanobots carrying various agents may help to accomplish the ultimate goal of cancer research: the eradication of solid tumors and vascularized metastases."[11]

But how do these nanobots move around inside your body? It's one thing to inject a fleet of nanosized robots into the bloodstream; it's another thing to direct them to the right locations. At Purdue University researchers have pioneered using ultrasound and magnetic fields to steer these tiny robots. This can power the nanomotors and also guide them throughout the body, where they can do everything from battling cancer to delivering drugs or mapping the human brain.

"What is neat about our design is that we have come up with a way to tailor the magnetic properties of the robot that will cause it to tumble in different ways when subjected to a rotating magnetic field," says David Cappelleri, a professor at Purdue University. "This allows it to tumble over different types of rough, bumpy, and sticky surfaces in both dry and wet conditions."[12]

This may remind some of you of the 1966 classic *Fantastic Voyage*, where a submarine crew is shrunk to microscopic size and ventures into the body of an injured scientist to repair damage to his brain. However, it's on an even smaller scale and doesn't require shrinking down humans.

Eventually, all of us may have fleets of nanoscale robots tumbling through our bloodstreams and battling hostile bacteria, viruses, and cancerous cells. These nanobots may also work to repair damaged tissues, kill parasites, and regulate our hormone levels. At the same time, we may have nanosensors strategically placed throughout our bodies to monitor our health in real time and alert our medical providers should any problems arise. Nanotech may even make it possible to collect in-depth data on our brains by converting mental activity into frequencies of light that can be registered by external sensors.

All this sounds fantastic, but there's a darker side to nanotech. Any machines that can eat away plaques, destroy cancer cells, and kill off bacteria can also potentially go awry and begin devouring our bodies. This hypothetical scenario has led to the gray goo theory. This is a catastrophic scenario where self-replicating nanomachines spin out of control and consume all biomass on Earth, turning every living thing into a gray mush.

If you want to read a sci-fi version of this, there's a thriller called *Plague Year* by Jeff Carlson, where a nanotechnology contagion devours all warm-blooded organisms living below ten thousand feet. This forces what remains of humanity to flee to higher elevations, where they struggle to survive.

But enough with the killer bots. If nanotech doesn't kill us, it will make us stronger. It will not only expand the limits of the human life span but help free us from worrying about many of the diseases and issues that our bodies have battled for millennia. By mastering the building blocks of the material world at their most fundamental level, we are in the process of reinventing everything around us. There may even come a time when we grow so dependent on this technology that we can't imagine life without it.

There is no end to our curiosity. Human expansionism will continue to drive us deeper down into the subatomic realm and farther out into the uncharted cosmos. The further we venture in either direction, the more we'll realize how vast our universe truly is.

MASTERS OF THE UNIVERSE OR MASS EXTINCTION?

Our insatiable curiosity and desire to push the boundaries of the known world is a two-edged sword, which has ensured human dominance but may eventually lead to our downfall. There's no denying that our natural intelligence combined with our drive to expand our knowledge and spread our DNA is what has made us the most successful animal on Earth, but it didn't have to be that way. There were several times in prehistory and more recently when *Homo sapiens* were on the brink of an extinction event.

The first event occurred 1.2 million years ago. It's unclear what caused this near extinction, but hypotheses include a giant meteor slamming into Earth or, more likely, changes in sea level and ocean anoxia triggered by global cooling or oceanic volcanism. What we do know is that the entire combined populations of *Homo sapiens*, *Homo ergaster*, and *Homo erectus* were reduced to fewer than thirty thousand, with an estimated breeding population of just ten thousand. This is smaller than the number of gorillas today. In other words, our ancestors were an endangered species.

But this initial setback didn't stop us. *Homo sapiens* bounced back and, once again, began to expand out from Africa to other parts of the planet. But 195,000 years ago, climate change caused the temperature around the world to plummet, and we entered an ice age. This took a catastrophic toll on *Homo sapiens*. Experts believe that the human breeding population shrank to just six hundred people scattered throughout the world.

Humanity eventually began to recover, until yet another cataclysm came our way. Roughly 70,000 years ago, Sumatra blew apart in a cataclysmic volcanic eruption, blanketing ash across much of the Earth and blocking out sunlight. This resulted in another ice age, and yet again, we came to the verge of extinction. Some scientists believe there may have been as few as a thousand humans left on the planet at that time.

It's hard to imagine a world where the entire human race could fit in a village. Today, with close to eight billion people spread across continents, it's difficult to go anywhere and not find someone living there. We have colonized nearly every bit of this planet, and even with the threat of climate change, we will probably manage without ever again coming so close to annihilation. Or will we?

Despite all of our technical achievements, our species' place on Earth isn't as secure as many people imagine. Given enough time, it's probable we will face another cataclysmic event that could put an end to our reign, and that's why colonizing Mars might not be such a crazy idea.

A lot of people have criticized the idea of spending billions to explore space when we have so many problems here on Earth that need to be addressed. They argue that the same money and resources should be spent on reducing greenhouse gas emissions, eliminating hunger and disease, and raising up the 1.3 billion people who live in extreme poverty. But is the threat of human extinction so small that we can safely ignore it? If you look at our recent past, you can see how close we've come to losing a large portion of the world's population.

As recently as 1995, Russian systems mistook a Norwegian weather rocket for a possible nuclear attack. Boris Yeltsin, then president of Russia, retrieved the launch codes and had the nuclear suitcase open in front of him. Fortunately, he did not enter the codes.

Twelve years earlier, in 1983, a Soviet system mistakenly reported that half a dozen intercontinental ballistic missiles had been launched from the United States and were headed toward Russia. Because there was so little advanced notice, the entire decision came down to a lone lieutenant colonel in the Soviet Air Defense Forces. If he had not decided to disobey orders and defy Soviet military protocol, we would have suffered a catastrophic nuclear war.

In 1962 the Cuban Missile Crisis nearly plunged the world into an all-out nuclear confrontation. It may seem like the world has changed since the Cold War ended, but we're actually more at risk now than ever. Nuclear proliferation has put weapons of mass destruction into the hands of more unstable nations, like North Korea, Pakistan, and Iran. There's also the threat of terrorists using nuclear weapons or hacking into a nuclear state's computer system and taking control of the missiles. Another possibility is a cyberhacker simulating a nuclear attack and triggering an all-out war.

There's also the risk of climate change. Even the most advanced climatologists don't fully understand what will happen as the Earth heats up. What if the melting of the ice caps accelerates further or the Gulf Stream is disrupted?

There's also the possibility of someone accidentally or intentionally releasing a genetically modified organism into the wild that winds up infecting and killing off large swaths of the world's animal or plant life. And then there's the risk of another superbug—a repeat of the Spanish influenza pandemic, which killed around fifty million people in 1918.

This doesn't even account for the most horrifying threats, which include those that come from beyond our planet. Scientists estimate that the asteroid that helped wipe out the dinosaurs was only seven to eight miles wide.

"The prospect of our planet experiencing a devastating, life-destroying impact by a comet or an asteroid may sound highly unlikely, but it is something that is almost certainly going to happen at some point in the future," says Natalie Starkey, a cosmochemist and author. "The question is when?"[13]

We've already identified thousands of asteroids out there large enough to destroy half the United States and send the world into darkness, but NASA scientists believe we're safe for at least the next one hundred years. That assumes they didn't miss any. Just to make certain, NASA is launching a new telescope into Earth's orbit to hunt these killer rocks. That said, stopping or diverting an asteroid on a collision course for our planet may not be possible.

What's a bigger unknown are solar flares. If a solar flare, like the one that hit the Earth 150 years ago, struck today, it would knock out all our satellite communications, electrical grids, and the internet. Much larger and more devastating solar flares are possible. One of these could release enough ultraviolet radiation and high-energy charged particles to destroy our ozone layer, cause mass DNA mutations, and disrupt all of Earth's ecosystems. Would we survive? No one knows.

"I'm not lying awake in bed at night worrying about solar superflares," says Gregory Laughlin, professor of astronomy and astrophysics at Yale

University. "But that doesn't mean that someone shouldn't be worrying about it."[14]

I haven't even touched upon the eruption of supervolcanoes, like the one that lies beneath Yellowstone National Park, or a geomagnetic reversal in which the Earth's magnetic field weakens and exposes our planet to solar winds—not to mention the threats posed by genetically modified organisms and other emerging technologies. Given that 99 percent of all species on Earth have gone extinct and modern humans have only been around for a couple hundred thousand years, our odds of lasting as long as the dinosaurs aren't as high as we may think they are.

"Either we spread Earth to other planets, or we risk going extinct," says Elon Musk. "An extinction event is inevitable, and we're increasingly doing ourselves in."[15]

This is some of what keeps Musk up at night and has driven him to launch SpaceX. He said he plans to send up to one million people to Mars by 2050 by launching three Starship rockets every day. Never one to aim small, he is determined to save the human race, whether we want to be saved or not.

MISSION TO MARS

SpaceX has a big, bold vision for colonizing Mars. At every turn, the company seems to be defying naysayers. It has announced plans to construct a hundred Starships per year and send a hundred thousand people from Earth to Mars every time the planets' orbits line up. Ultimately, it envisions a thousand Starships shuttling people and supplies back and forth.

This might sound like an impossible feat, but it's actually the easy part. The hard part begins once humans arrive at the red planet. There are so many unanswered questions that still need to be resolved: How will people safely land on Mars? Once they do land, will they ever be able to get off, or are they trapped for life? And can anyone hope to survive, let alone thrive, in the harsh Martian atmosphere?

Musk has plenty of critics. In fact, the majority of experts believe his plans are overly overambitious, and he's relying too much on faith instead of facts. But this is precisely what makes Musk such an exceptional entrepreneur. He's willing to bet everything on a belief, even when everyone around him doubts the feasibility.

SpaceX may have the Starships to take us to Mars, but landing on the planet, especially with fragile payloads, like human beings, isn't going to be easy. There's also the ongoing question of supplying the colonists with everything necessary to sustain life. If there's any sort of screw up, the colonists would have to wait another twenty-six months for Earth and Mars to align properly for interplanetary missions.

Around 3.8 billion years ago, Mars lost its magnetic field, which could have protected life on the planet from the harsh cosmic ray particles that tear through cells, cause cancers, damage brains, and shred nervous tissue. "With the current shielding technology, it is difficult to protect astronauts from the adverse effects of heavy ion radiation. Although there may be a way to use medicines to counter these effects, no such agent has been developed yet," says Kamal Datta, a project leader at NASA.[16]

This means that anyone staying on the planet will need to be protected from this radiation, if they hope to survive long. Unless we can come up with special materials to shield people from the environment, it's not going to be much fun for the colonists. But radiation isn't the worst of it. The lack of a magnetic field has meant that the Martian atmosphere started leaking into space, making it increasingly inhospitable.

"If you are unprotected on Mars . . . your blood would boil, even at ambient temperature," explains Pascal Lee, chairman of the Mars Institute. "Like popping a can of Coke, you would fizz to death."

While Earth's atmosphere is 78 percent nitrogen and 21 percent oxygen, with trace amounts of water vapor, carbon dioxide, and other gases, the Martian atmosphere is 95 percent carbon dioxide. "We need to breathe oxygen," says Lee. "There's no free oxygen in the Martian atmosphere. You cannot breathe this gas. You would die of hypoxia within minutes."[17]

Then there's the temperature. At the Martian equator, it can reach as high as 70 degrees Fahrenheit, but the nighttime temperatures are lethal. On average, the temperature plunges to about minus 80 degrees Fahrenheit. Without the right spacesuits and living quarters, people won't last long.

If that isn't enough to deter you from immigrating, Martian toxic dust is so fine and abrasive that it would wreak havoc on human lungs. A few weeks of breathing this stuff, and you'd be in your grave. Clearly, Mars is no tropical paradise.

This is only part of the problem. Anyone colonizing Mars will need safe living structures, as well as clean air, water, and food, and a source of power to keep everything running. Initially, at least, all of this would have to be imported, which would be incredibly expensive. Who is going to pay for this? SpaceX has raised a sizable amount of capital, but how much will this cost, and what will investors get in return? Will they have to wait a hundred years or more to see a payoff? Most venture funds have a six to twelve-year horizon.

When it comes to growing crops, Martian soil is devoid of essential nutrients, laced with nasty chemicals, and so fine-grained that any water would likely seep through. This means everything will have to be grown in specially designed greenhouses with imported soil or hydroponics. Add this to the list of expenses.

There's a lot of talk about terraforming Mars, but even if that's possible, it won't happen overnight. Most scientists believe it would take hundreds of years before the thin Martian atmosphere could be transformed into anything close to that of Earth's.

The weak gravity on Mars will also impact human health. It may increase the risk of kidney stones and bone fractures. People may also suffer from a loss of muscle mass, strength, and endurance. This could affect their hearts, leading to diminished cardiac function. In addition, microgravity interferes with our ability to maintain balance, stabilize vision, and understand body orientation. It's easy to forget that our bodies evolved on Earth and are calibrated to the precise conditions of our home planet. Take us somewhere else, and we're like fish out of water.

Despite all these impediments, SpaceX is charging ahead undaunted. You have to admire the chutzpah. Many people, including myself, believe we should be focused on sending robots, not humans, to the red planet for the next fifty years or so. Robots won't have all these issues, and they can prepare the way for later human habitation.

But Musk doesn't want to wait. He's both incredibly optimistic and fearful that an irreversible extinction event may come sooner than expected.

For Musk, Mars is humanity's backup drive, and he wants it operational before the big crash comes.

BUILDING AN INTERPLANETARY ECOSYSTEM

Space exploration is a growing industry that includes hundreds of companies, as well as the world's most powerful government space agencies. Ultimately, the health of the entire space ecosystem is going to get humans to Mars and beyond.

NASA wants to put humans on Mars and sees the Moon as its stepping-stone. The idea is to build a lunar base and use it to springboard to Mars. The space agency is collaborating with SpaceX, Blue Origin, and Boeing to make this happen.

"Fifty-five years after President Kennedy challenged the nation to put a man on the Moon, the Senate is challenging NASA to put humans on Mars," said Bill Nelson, after approving funding for a Mars program in the United States Senate. "The priorities that we've laid out for NASA in this bill mark the beginning of a new era of American spaceflight."[18]

Armed with a bigger budget and bolder vision, NASA isn't stopping there. On the surface of Mars, the InSight lander is listening for quakes, and in 2021 was joined by the Mars rover Perseverance. New Horizons is exploring the Kuiper Belt, a region containing millions of chunks of ice left over from the solar system's birth. NASA also sent off a probe to rocket past the sun twenty-four times and gather data. And it plans to fly a nuclear-powered helicopter over the surface of Titan, Saturn's icy moon, to scan for signs of alien life.

Inspired by Silicon Valley, NASA now has another moonshot program. Some of the far-out ideas include genetically engineering new organisms that can survive in the extreme Martian environment and sending them to the red planet to detoxify and fertilize the soil; equipping a craft with an ionic thruster, attaching solar panels to it, and then blasting it with a powerful laser to create a supercharged solar sailing ship; sending squishy robots, which won't get mired in debris or stuck in cracks, to flop about on asteroids and gather samples; and launching an entirely mechanical probe, like something out of a steampunk novel, that contains no electronic parts that would get fried on Venus' high-pressure, infernal planetary surface.

Another one of NASA's out-of-the-box ideas is to move its manufacturing into space. It has already awarded the start-up Made In Space a contract to demonstrate 3D-printing spacecraft parts while in orbit. This would save NASA from having to launch large man-made structures into orbit, which is expensive and inefficient. "In-space robotic manufacturing and assembly are unquestionable game-changers and fundamental capabilities for future space exploration," says Jim Reuter, an administrator at NASA.[19]

But these 3D printers would still need raw materials. That's where Gerard O'Neill, a Princeton University professor and space visionary, comes in. In 1974 he proposed using an electromagnetic railgun to shoot payloads of ore mined from the Moon into Earth's orbit. This just might become possible, enabling a 3D printing factory orbiting the Earth to harvest the raw materials.

One of SpaceX's biggest competitors is Blue Origin. With Jeff Bezos at its helm, the start-up doesn't lack resources, but it has taken a more modest approach. Instead of aiming for Mars, Bezos has set his eyes on the Moon. The company's motto sums up its approach: *gradatim ferociter*, which is Latin for "step by step, ferociously."[20] Blue Origin has stated that its mission is to develop "technologies to enable human access to space at a dramatically lower cost and increased reliability."[21]

Bezos wants to see people living and working in space, but his goal isn't necessarily to colonize other planets. Instead, his vision is to get heavy industry and power generation off our planet before we completely destroy the environment. He's looking at the moon as the perfect place to relocate our most polluting industries.

"We should build permanent settlements on the Moon's poles where we can get water and solar power," says Bezos. "We know things about the Moon we didn't know back in the 1960s and 1970s, and with reusable rockets we can do it affordably."[22]

Robert Bigelow, another billionaire, is investing his fortune in building giant inflatable space habitats. Having made his money in the hotel business, Bigelow is taking his hospitality expertise to outer space. Similar to an industrial-strength balloon, these habitats will be stuffed into rockets and blasted into orbit. Once there, the habitats can expand outward, creating living and working quarters for astronauts and space tourists.

Bigelow envisions his balloon-like hotels being used throughout the solar system and even in deep space. Not to be outdone, NASA is testing out its own inflatable habitats for the Moon and Mars. The US space agency has designed these structures using sturdy fabrics, rubberlike coatings, and other materials that can protect astronauts from harsh temperatures, radiation, and high-velocity rocks and debris.

Scientists are also looking into growing buildings in space. Taking materials from Earth into space is extremely expensive, so NASA is exploring how to grow structures from mushrooms. Its researchers have proposed creating a habitat made from lightweight materials embedded with fungi. Once the habitat arrives on the Moon or Mars, all the astronauts need to do is add water and the fungi will do the rest.

"Right now, traditional habitat designs for Mars are like a turtle, carrying our homes with us on our backs, a reliable plan, but with huge energy costs," says astrobiologist Lynn Rothschild, the principal investigator of the proposal at NASA. "Instead, we can harness mycelia to grow these habitats ourselves when we get there."[23]

Like so many other technologies, these fungal bricks may even find their way back to Earth. NASA claims that mycelial materials, already commercially produced, are known insulators, fire retardants, and do not produce toxic gases. Metrics for these materials show compression strengths superior to lumber, flexural strength better than reinforced concrete, and competitive insulation values. This means all of us may be living in mushroom homes someday.

If you're thinking that it would be nice to have a summer fungus cottage on the Moon, but the lack of oxygen may cut your vacation short, don't worry. The European Space Agency has that covered. Its researchers have found a way to produce oxygen from lunar dust. Surprisingly, lunar dust contains between 40 and 45 percent oxygen by weight. "Being able to acquire oxygen from resources found on the Moon would obviously be hugely useful for future lunar settlers, both for breathing and in the local production of rocket fuel," says Beth Lomax of the University of Glasgow.[24]

All they have to do is mix the lunar dust with molten calcium chloride salt, heat the mixture up to 950 degrees Celsius, and run a current through it, releasing the oxygen. If it works on a large scale, we can all breathe a sigh of relief.

Not only is there oxygen on the Moon, but both the Moon and Mars have a substantial amount of water stored in their icecaps. This is good news for potential colonists. NASA even speculates that 3.5 billion years ago, before Mars lost its atmosphere, the planet was habitable.

"The detection of organic molecules and methane on Mars has far-ranging implications in the light of potential past life on Mars," wrote astrobiologist Inge Loes ten Kate. "Curiosity [a rover sent in 2011 as part of NASA's Mars Science Laboratory mission] has shown that Gale crater was habitable around 3.5 billion years ago, with conditions comparable to those on the early Earth, where life evolved around that time."[25]

Even more promising, there is running water on Mars.

"There is liquid water today on the surface of Mars," says Michael Meyer, the lead scientist on NASA's Mars exploration program. "Because of this, we suspect that it is at least possible to have a habitable environment today."[26]

In the summer months on Mars, water runs down canyons and crater walls. Researchers are unsure where this water comes from, but it may rise up from aquifers or melting ice underground or condense out of the thin Martian atmosphere. Roberto Orosei, a professor at the University of Bologna, has even detected a giant lake, 12.4 miles wide, below the surface of Mars. This means that colonists could potentially access enough water not only to keep them alive but also to irrigate farms and produce hydrogen fuel through electrolysis.

Speaking of farming, a team of Harvard, Caltech, and University of Edinburgh scientists believe a layer of aerogel just two to three centimeters thick may be enough to create gooey greenhouses that protect plants from the extreme Martian temperatures and solar radiation. They've re-created Mars-like conditions in their lab and have been testing out the aerogel.

"It's a very different approach to a lot of these global terraforming ideas," says Robin Wordsworth, a researcher at Harvard. "But it has the advantage that we could actually do it in the next few decades rather than much further in the future."[27]

That sounds great, if colonists don't mind being vegetarians. But what about a more protein-rich diet? Raising livestock on Mars isn't going to be easy, and importing enough meat to feed a million people is out of the question. So what other options are there? Researchers at the University of Central Florida think they have the answer: bug farms.

"If you want to feed a large population on another planet, you have to move away from the idea of watery vegetables and really think about the tremendous amounts of energy, water, and raw materials needed to produce enough calories," says Kevin Cannon, a planetary scientist and author of the study. "Bugs and bioreactors are the way to go, if people can get over the gross factor."[28]

If you don't mind eating cockroaches, putting up with a few health problems, and living inside a giant mushroom, there may be a job waiting for you. With plenty of water on both the Moon and Mars and a lot of human ingenuity, it seems like the Mars-or-bust crowd might not be so crazy after all. We may see colonies of people living and working on the red planet in our lifetimes.

SPACE PIONEERS AND THE GOLD RUSH

Space is the Wild West right now, and entrepreneurs are rushing to stake out their claims and get rich quick. Nothing epitomizes this more than the motley band of asteroid hunters.

It may sound like something out of a comic book, but Goldman Sachs is bullish on the concept. America's largest investment bank estimates asteroid mining could become a trillion-dollar industry. A single small asteroid of a hundred feet in diameter could contain enough precious metals to be worth $50 billion. Each space rock is literally a flying goldmine that can contain all sorts of valuable minerals, ranging from platinum and rhodium to iridium. The mother lode would be a three-thousand-foot-wide asteroid, which could contain trillions of dollars' worth of platinum. This makes the renowned Comstock Lode look paltry, considering it would be worth a mere $670 million today.

Noah Poponak, an aerospace and materials analyst at Goldman Sachs, wrote, "Prospecting probes can likely be built for tens of millions of dollars each, and Caltech has suggested an asteroid-grabbing spacecraft could cost $2.6 billion."[29]

This is enough to make some entrepreneurs want to grab their picks and shovels and head off to space in search of riches. Several start-ups have already put up their shingles to do exactly that. Deep Space Industries was one of them.

"Crazy ideas: that's what moves culture forward," says Rick Tumlinson, cofounder of Deep Space Industries. "Nothing says this is impossible, except our own belief systems."[30]

Planetary Resources, another asteroid mining start-up, raised $50 million, of which $21 million came from big-name investors, including Google's Eric Schmidt and filmmaker James Cameron. The biggest potential in asteroid mining isn't necessarily bringing the precious metals back to Earth and selling them. In fact, too much precious metal flooding the market could cause prices to plummet, making the whole endeavor much less profitable. The real long-term value in asteroid mining may lie in harvesting materials for use in outer space.

The ice on asteroids could potentially serve as a rocket propellant, while the metals might be used to build everything from space stations and giant antennas to solar farms. Using the materials in outer space could play a vital role in building the space economy, while eliminating the cost of bringing asteroids all the way back to Earth.

"There was a lot of excitement and tangible feeling around all of these things that we've been dreaming about," says Chad Anderson, CEO of Space Angels, a venture capital fund. "Both companies were really good at storytelling and marketing and facilitating this momentum around a vision that their technology never really substantiated."[31]

But neither start-up has delivered on their promises. Deep Space Industries wound up selling out to Bradford Space, while Planetary Resources was acquired by ConsenSys, a blockchain software company.

"I was shocked. I think they wanted to acquire the equipment and assets," says Anderson Tan, an early investor in Planetary Resources. "For what? I'm not so sure."[32]

"The bottom line is that space is hard," says Henry Hertzfeld, the director of the Space Policy Institute at George Washington University. "It's risky, it's expensive, lots of high up-front costs, and you need money. You can get just so much money for so long."[33]

The main problem was that these start-ups were too early. The ecosystem needs to develop before asteroid mining becomes feasible. "If you mine an asteroid, most likely you'll send it to the moon to process it. It wouldn't be processed on Earth because the cost would be tremendous," says Tan. "So, then it's like a chicken-and-egg problem: Do we mine first and then develop a moon base, or invest in building up the moon and then go to asteroid mining?"[34]

It looks like asteroid miners need to wait for Blue Origin, NASA, and others to lay the groundwork. Then they can swing back into action. In the meantime, there are plenty of other wild opportunities within reach. StartRocket, a Russian start-up, said it wants to display giant billboard-style advertisements in the night sky using arrays of CubeSats.

StartRocket isn't alone. Rocket Labs, a New Zealand start-up, shot a geodesic disco ball named the Humanity Star into orbit in 2018. The company says the bright, blinking satellite was "designed to encourage everyone to look up and consider our place in the universe."[35] (The Humanity Star met an earlier demise than expected, deorbiting and burning up in the atmosphere only a few weeks after its launch.)

If that's not enough to light your fire, Japanese start-up ALE wants to deliver the world's first artificial meteor shower. Each satellite would carry four hundred tiny balls that when ejected from the spacecraft will glow brilliantly as they plunge through the atmosphere. The business model is to sell this service to anyone who will pay for it. This includes billionaire birthday parties.

This may be the first ugly step toward transforming our pristine night skies into a commercial zone. HyperSciences wants to make sending satellites into orbit cheap enough for all sorts of smaller businesses to jump into the game. It has taken a ram accelerator system and flipped it on its head, so that it can fire projectiles upward at nine times the speed of sound. The goal is to shoot small payloads into space out of a giant tube at a much lower cost than conventional rockets.

SpaceX, which is in the process of deploying thousands of satellites in a megaconstellation known as Starlink, has come under fire. Although its goal of providing fast, reliable internet service to people around the world with little or no connectivity is a worthy one, how the company is going about it has ruffled feathers. Many scientists claim this is an existential threat to ground-based astronomy, not to mention an annoyance to anyone who enjoys looking up at the stars.

"What surprised everyone, the astronomy community and SpaceX, was how bright the satellites are," says Patrick Seitzer, an astronomer at the University of Michigan.[36]

SpaceX is working on different coatings to make the satellites less visible, but their brightness is only one problem. There's also the issue of floating space junk. Today, there are 128 million pieces of space junk larger than one millimeter and 34,000 pieces larger than ten centimeters, along with 3,000 dead satellites in Earth's orbit. These can wreak havoc on anything sent into space, tearing holes in space stations and damaging satellites and rockets.

When two satellites collide, they can smash apart into thousands of pieces, creating ever more debris. In 1978 Donald Kessler pointed out that too much space junk in orbit could result in a chain reaction where collisions increase exponentially. As more companies pile into the space economy, there needs to be regulation around what is sent into orbit and how it behaves.

All these problems can be solved. It just requires a degree of foresight and cooperation between governments. If the satellites are put into decaying orbits, they will burn up in the atmosphere after a time, removing the threat. Modern satellites also have the ability to detect and avoid space debris. These are some of the solutions that can mitigate any future problems, as long as companies act responsibly.

The space pioneers don't want governments to overreact and impose so much regulation on space exploration that it cripples the industry just as it's about to take off. The room for innovation is enormous, and most experts agree that private industry working in collaboration with governments is our best bet for the future.

THE SPACE ECONOMY

Billions of dollars in venture capital have been fueling the growth of the space industry. While US companies have led the way, Chinese companies took in roughly one-third of investment dollars. Most funding has gone to highfliers, like SpaceX, Blue Origin, and Virgin Galactic, but investment in smaller start-ups has also increased. The US Chamber of Commerce expects the space economy to reach $1.5 trillion by 2040.

Satellite ride-sharing has lowered costs significantly, allowing more players to enter the market and putting more satellites into orbit. There are now hundreds of start-ups focused on building and launching satellite-based services. There's also a growing space tourism industry, with SpaceX, Blue Origin, and Virgin Galactic all offering rides into orbit. Most trips will only offer a few precious minutes for passengers to unstrap themselves and float around; then it's time to head back to Earth. If you have enough money, you may even be able to book a jaunt around the moon. After this will come floating space hotels and trips to Mars. Space flights won't come cheap. However, like commercial air travel before it, prices will steadily decrease as economies of scale kick in.

If you want to save some money, you may want to try going up in a space balloon. Two start-ups, Zero 2 Infinity and World View, are offering rides in space balloons for less than half the price of Virgin Galactic. Of course, you get what you pay for. These space balloons will only reach altitudes of about twenty-two miles, whereas space technically begins at the Kármán line sixty-two miles above sea level. That said, passengers will get a nice view, including the sun rising over the curvature of the planet.

If you plan on going farther than Earth's orbit or the Moon, spaceflight can take a long time. A journey to Mars can range from 150 to 300 days, depending on the speed of the starship and the alignment of the planets. I get antsy on twelve-hour flights. I can't imagine sitting cooped up in a tin can with other passengers for months on end.

This is what's driving the development of artificial emotional intelligence at NASA. Tom Soderstrom, CTO of the Jet Propulsion Laboratory, is working on an AI that can provide emotional support for astronauts on deep-space missions. "We want to have an intelligent assistant that can control the spacecraft's temperature and direction, figure out any technical problems—that is also watching human behavior," says Soderstrom.[37]

An emotional AI would learn to read each individual's facial expressions, body language, and tone of voice, then use this data to foster collaboration between passengers. It may even detect stressful situations and work to defuse them before they become issues. This means that you may be spending a lot of time on your flight to Mars interacting with an AI assistant who is continuously monitoring your psychology and physical state, while working behind the scenes to make your journey both pleasant and productive.

What people overlook sometimes is that the space economy isn't just about the big, sexy things like SpaceX's Big Falcon Rocket, Dragon cargo ships, and Starships—not to mention the plethora of space stations, satellites, and giant telescopes. The ecosystem also has a less visible side. It's all the researchers, government agencies, and companies working to make the software, mechanical parts, instruments, suits, foods, materials, and a million other little things that will be necessary.

There's a huge opportunity for entrepreneurs in the less visible but potentially more lucrative field of producing the picks and shovels of interplanetary commerce and communication. The more progress pioneers like SpaceX and Blue Origin make, the bigger the pie for everyone.

WARP DRIVE: FAR-OUT IDEAS

What's coming next? Will it ever be like *Star Trek*, where we have warp drives, tricorders, and teleportation? You might be surprised at some of the wild ideas scientists and entrepreneurs are dreaming up.

A big problem with using conventional starships is that it can take a long time to reach distant planets. It would take 2.5 million light-years to reach Andromeda, the nearest spiral galaxy to ours. Even traveling to Mars is daunting. That said, there may be a solution. How about sleeping through the entire trip?

SpaceWorks, an Atlanta-based start-up, received two rounds of funding from NASA to investigate the feasibility of building a hibernation chamber. Just like bears crawl into a cave for the winter to hibernate, researchers believe we can lower the body's core temperature by around 48 degrees Fahrenheit, inducing a sleep state called torpor. A 2009 report showed how a patient at the Mayo Clinic remained in a torpor state for two weeks.

Even for a relatively short trip within our solar system, this could make a big difference. Going hypothermic would reduce the body's metabolic rate by as much as 50 to 70 percent, meaning less consumption of oxygen, food, water, and other scarce resources. Hibernation has the added benefit of helping to counter microgravity's effects on the body, including bone demineralization, intercranial pressure, and muscle atrophy. Bears don't lose muscle over the winter. It also would allow many more people to be crammed like sardines into a single starship.

Someday, people with Alzheimer's or cancer may choose to hibernate until scientists develop cures for their diseases. I can also see this as being popular with billionaires, who want to wait it out until someone comes up with a way to reverse aging. Or people may want to try it out so they can pop up again fifty years from now and experience what the world is like in the future.

"I was watching *Star Trek*," says Miguel Alcubierre, director of the Nuclear Sciences Institute at the National Autonomous University of Mexico. "It got me to thinking if there was any way in which you could come up with a geometry of space-time that was similar to this idea of warp drive in science fiction that allowed you to travel faster than light."[38]

He wound up coming up with the concept of a warp bubble that would expand the space behind the starship while compressing the space in front. This could theoretically enable the ship to travel faster than the speed of light, even though it isn't moving at all. Since the theory of relativity does not impose any speed limits on the expansion of space, this would not defy the laws of physics. The one problem remaining to be solved is that distorting space-time requires matter with an energy density less than zero. Unfortunately, this is not yet known to exist. That hasn't dissuaded Harold White, a NASA physicist. According to the *New York Times*, White and his team of scientists claim to "have found a way to configure the hypothetical negative energy matter so that the warping could be accomplished with a mass equivalent to the Voyager spacecraft."[39]

"What this does is it moves the idea from the category of completely impossible to maybe plausible," says White. "It doesn't say anything about feasible. And so, unfortunately, that point usually gets missed a lot."

On a more practical level, a start-up in Seattle called Ultra Safe Nuclear Technologies wants to build a nuclear thermal propulsion engine that could cut the travel time to Mars in half. This isn't exactly warp speed, but a three-month trip to Mars would make a big difference in colonization efforts. NASA is evaluating the technology now to determine its feasibility.

A slightly more down-to-earth idea is a space elevator: Instead of hauling everything into orbit on rockets, why not build an elevator-like transportation system? Researchers at Japan's Shizuoka University have already begun to test the feasibility. If it works, it could dramatically lower the cost of moving large amounts of materials and equipment into near-Earth orbit. A space elevator involves a vehicle that can climb a cable running from the Earth all the way up into space. The cable could be made from a number of materials, including extremely lightweight but tough carbon nanotubes or diamond nanothreads.

Scientists at Cambridge and Columbia Universities have proposed anchoring a superstrong, thin cable (a few millimeters in diameter) to the Moon's surface that would extend into geostationary orbit about 159,400 miles above our planet. Once this cable is in place, they envision advanced machines climbing the cable as they lift payloads into orbit. Scientists are still debating whether any sort of space elevator is practical. Many believe that carbon nanotubes and other materials simply won't work. Time will tell who's right.

Even better than hibernating like a bear would be teleporting to Mars. What most people don't realize is that teleportation *is* possible. No one has figured out yet how to teleport physical objects, let alone humans, from one place to another, but we can teleport *information* about photons and even molecules through a process called quantum entanglement.

Albert Einstein called this "spooky action at a distance" because it defies all conventional logic. When two particles become entangled at a quantum level, they can be separated by huge distances, and yet the description of any changes in the quantum state of one particle will be instantly reflected in the other. In other words, it seems to defy Einstein's theory of relativity where nothing can travel faster than the speed of light. As you can imagine, this didn't make Einstein happy.

Back in 2017 researchers in China successfully teleported information between a photon on Earth and its entangled pair on a satellite orbiting 311 miles away. In 2019 European researchers performed chip-to-chip quantum teleportation.

"We were able to demonstrate a high-quality entanglement link across two chips in the lab, where photons on either chip share a single quantum state," says Dan Llewellyn, a researcher at the University of Bristol.[40]

These are the first steps toward building a quantum internet in space. Back on Earth, the quantum internet is already being tested in some cities, like Beijing, Chicago, and New York. As the quantum internet matures, it will enable a highly secure form of global and possibly interplanetary communication. It could also provide exponentially more bandwidth, enabling quantum supercomputers to run extremely complex simulations in the cloud. Distant clocks could be synchronized about a thousand times more precisely than today's atomic clocks. This would improve GPS systems and the mapping of Earth's gravitational field. It could also lead to better optical and radio telescopes.

None of this will get you to Mars in the blink of an eye, but it will dramatically improve communications, internet security, and our ability to understand the cosmos.

LIFE ON OTHER PLANETS

With at least 200 billion galaxies in the universe, the odds are that intelligent life exists on other planets. It's one thing to find some microbes on Mars but another to discover conscious beings that evolved separately from us. The question is, How far away are these alien life-forms from Earth and can we communicate with them? The universe is so vast that the chances of us bumping into one another are slim, right? Not necessarily. A recent study estimates that dozens of intelligent alien civilizations are hiding in our galaxy right now—all of them capable of communicating.

"There should be at least a few dozen active civilizations in our Galaxy under the assumption that it takes 5 billion years for intelligent life to form on other planets," says Christopher Conselice, professor of astrophysics at the University of Nottingham.[41]

Some believe we may already have been visited by aliens. No, this is not a conspiracy theory. British astronaut and astrobiologist Helen Sharman suspects invisible aliens might be living among us right now. These aren't aliens like you'd see in the movies but unintelligent microscopic organisms existing in a shadow biosphere. "By that, I don't mean a ghost realm, but undiscovered creatures probably with a different biochemistry," writes Sharman. "This means we can't study or even notice them because they are outside of our comprehension."

This isn't beyond the realm of possibility. While all terrestrial life is carbon based, silicon is chemically similar to carbon and is abundant on our planet, yet it is surprisingly unutilized by biological organisms. But in 2016 researchers at Caltech managed to coax living cells to bond with silicon. "We do have evidence for life-forming, carbon-based molecules having arrived on Earth on meteorites," Sharman says, "so the evidence certainly doesn't rule out the same possibility for more unfamiliar lifeforms."[42]

In our lifetimes, the odds are that we won't have any close encounters of the third kind, like in the eponymous movie, but we will probably live to see new forms of life thriving on other planets. Using gene-editing technologies, we're already developing a host of genetically modified plants, animals, and microbes that can adapt to our needs here on Earth. There's no reason we can't design these for alien environments, like those on Mars, Titan, or Enceladus. These life-forms may have never existed before, and they may be so different that they cannot survive back on Earth.

The radiation levels on spacecraft and alien planets can be deadly—not to mention the environmental toxins and extreme temperatures, which is why many scientists believe we must genetically modify most life-forms before sending them into deep space for long periods. "You can't send someone to another planet without genetically protecting them if you are able to," says Christopher Mason, a professor at Cornell University. "That would also be unethical."[43]

Speaking of new life-forms, geneticist George Church is pioneering the development of genetically modified animals. He's working to create the world's first "mammophant" by taking DNA from a woolly mammoth and injecting it into an elephant embryo. Will this bring back the woolly mammoth that went extinct 4,000 years ago?

"Our aim is to produce a hybrid elephant-mammoth embryo," says Church. "like an elephant with a number of mammoth traits."[44] Church's work could pave the way for designing animals better suited to survive on other planets.

Another one of Church's projects is to synthesize the complete human genome. His goal is to build, from scratch, all the genes that make humans human. This technology could be a critical next step toward spreading the seeds of humankind across the galaxy, especially when coupled with an artificial womb.

Researchers at the Children's Hospital of Philadelphia are developing this artificial womb right now and have already tested it with a lamb fetus. They're doing this to save the lives of prematurely born babies, but it could be adapted for use during space travel. Instead of carrying a baby to term for nine months, which may be dangerous in a low-gravity environment, colonists may opt for in vitro fertilization and artificial wombs as a safer alternative. Someday, we may have elaborate breeding facilities on Mars and other planets, which produce not only humans but also genetically modified pets and livestock.

Some astrobiologists want to take it a step further using synthetic human genomes. This would involve recoding human chromosomes to help us resist radiation damage, fend off viruses, and prevent muscle atrophy and osteoporosis in low-gravity environments. This research could also lead to developing prototrophic humans. These are people that can subsist on a restricted diet that lacks the nutrients we find on Earth. With advanced genetic engineering, we may enable our kidney cells to synthesize the amino acids, vitamins, and nutrients we need to survive long journeys with limited food.

"I don't want it said that I am making green people, and I am not suggesting we do this any time soon. But I am suggesting that if you want to do intergalactic travel, you need to solve the problem of being totally self-sufficient," says Harris Wang, a professor at Columbia University. "We are putting humans in very extreme conditions, and from that perspective this seems to be one idea for a long-term plan."[45]

Researchers have already learned how to manufacture embryos from stem cells. For long journeys to distant planets, we may not send humans or animals. Instead, we may only need to send stem cells and DNA, along with artificial wombs. Once the spacecraft arrives at a habitable destination, we can have robots take over, modifying the genes and breeding the necessary life-forms according to the environmental conditions. The nice thing is that DNA and stem cells are extremely compact, which means we could fit an entire Noah's Ark in a test tube.

At a certain point, what we call humans may actually be another species that can only survive on these planets. If these posthumans return to Earth, they may need protective suits to walk outside. Many of the animals and plants may become so alien that people back on Earth may not even recognize them.

Scott Solomon, an evolutionary biologist and professor at Rice University, points out, "Eventually, people living in space could evolve to be different enough from people on Earth that we would consider them to be different species."[46]

Instead of attempting to terraform Mars into Earth, we may transform ourselves into Martians, along with our plants, animals, insects, and whatever else we need to survive there. In other words, we may become the scary aliens we see in the movies.

If this sounds too much like science fiction, you should know that researchers at Cambridge University have created the world's first living organism with fully redesigned DNA. It's a microbe with a completely synthetic and radically altered DNA code that makes it resistant to viruses.

"They have taken the field of synthetic genomics to a new level, not only successfully building the largest ever synthetic genome to date, but also making the most coding changes to a genome so far," says Tom Ellis, a synthetic biologist at Imperial College London.[47]

All this sounds interesting, but there are ethical issues to consider. Not only do many people object to the idea of humans playing God, but we also have to think of the planets themselves. Once we arrive, we will inevitably introduce all sorts of bacteria, which could permanently alter them, making it difficult for scientists to determine which life-forms evolved on the planet and which were introduced from Earth.

A typical spacecraft can contain hundreds of thousands of bacterial spores. Those pose a real threat to the genetic purity of a virgin planet.

"Bacterial spores are very hardy," says Madhan Tirumalai, a microbiologist at the University of Houston. "They can remain in the environment for millions and millions of years until they find the right conditions to start germinating."[48]

When an Israeli lunar lander crashed on the Moon, it may have dumped a pile of tardigrades onto the surface. These eight-legged micro-animals are extremely hardy and can survive even in high-pressure, extreme-radiation environments like outer space. NASA's Office of Planetary Protection isn't too concerned about the Moon because it's void of life, but doing the same on Mars, which may harbor a number of native microorganisms, could be a disaster for scientists who want to study how life evolves on other planets. There's been a heated debate going on between those who want to preserve these planets in their pristine condition and those who want to colonize them.

Most likely, the colonists will win out. Once we begin landing large numbers of spacecraft on a planet, it will be virtually impossible not to bring along invasive microbes, and when humans arrive, it's game over. We are walking biocontamination vessels. Whether this is a moral catastrophe or a bold step forward for humankind depends on what you view as progress.

No matter what, it's not possible to stop human expansionism. It's one of the five fundamental forces driving the advancement of our civilization. So get ready, Martians, here we come!

FORCE 4
DEEP AUTOMATION

Deep Automation, the force driving humans to algorithmically automate all the underlying processes for managing, growing, and sustaining life, will accelerate innovation, create wealth, and free us from the need to work.

When most people think of automation, they imagine robots doing all the work. What's often overlooked is that without artificial intelligence, robots are just dumb machines. It's the intelligent algorithms that enable them to perform advanced tasks. AI, not machines, will be the major driver of automation moving forward. It will power everything from humanoid robots to the array of smart devices entering our homes, offices, factories, and bodies. Intelligent algorithms will also power the software used to manage our infrastructure, health care, finance, transportation, and government.

Over the coming years, the force of deep automation stands to upend our society. If we contrast our lives with those of our parents and grandparents, we can see how much has changed in such a short time. We've gone from a world dominated by pen and paper to a digital existence, where we walk around with computers in our pockets and can't imagine life without the internet. With new technologies emerging at an ever-faster rate, the coming wave of innovation will be significantly more transformative.

What we know for certain is that in the future, almost no job will remain unaffected. There will probably still be chefs, accountants, and teachers for the next decade or two, but what they do on the job and how they do it may look nothing like it does today. Technology will alter and take over much of the routine work people perform, while at the same time enabling humans to do things that were previously too difficult or unimaginable.

In this section, we will highlight the core developments driving deep automation. These include advances in everything from artificial narrow intelligence (narrow AI or ANI) and robotics to new types of sensors, computer networks, and the internet of things (IoT). Out of all these technologies, we will spend a disproportionate amount of time on narrow AI and its impact on automation because this has the most far-reaching implications. By narrow AI, we mean algorithms that are capable of performing clearly defined tasks, like guiding a robotic vacuum around a house, translating English into Swahili, or recognizing a dog in a photo.

All the AIs we see in the world today are narrow. None of them comes close to reproducing human consciousness. This is because artificial intelligence capable of understanding the world and learning like a human hasn't been invented yet. Computer scientists call it artificial general intelligence (AGI), but for those who are less technical, we'll call it superintelligence. We will address superintelligence in the final section of this book, where we'll explore both the practical and philosophical implications of creating machines capable of matching and surpassing the human brain.

Deep automation doesn't require superintelligence. This means it is achievable in the near term. In fact, most of the core technologies already exist and are being put to use in our factories, offices, and homes. Many of us experience ANI-powered automation in our everyday lives, from virtual assistants, like Siri and Alexa, to recommendations on Netflix and Amazon. It's something we've come to take for granted, but it also has unseen and less discussed social and economic implications, which stand to upend our carefully balanced social order.

For example, as we automate, will we be able to retain any sense of privacy and control over our lives? Given enough data, can algorithms predict everything from our buying habits to our romantic partners? How will humans and ever-more-capable computers work together symbiotically? What does it mean to automate our eldercare, children's education, and work environments? And finally, will we come to rely on our machines too much and lose an essential element of our humanity?

SMART CITIES: DREAMS OF MAGIC KINGDOMS

Countries like Saudi Arabia are building entire smart cities from the ground up. The goal is to automate as much as possible, using the latest technologies to make cities more efficient, productive, ecological, and intelligent.

Crown Prince Mohammed bin Salman is betting half a trillion dollars to turn barren desert and an undeveloped shoreline into a city of the future. The Saudis are calling this new metropolis Neom, and they want it to outshine Silicon Valley as an innovation hub. The plan includes flying taxis, robot maids, high-tech hospitals, Michelin-star eateries, a Jurassic Park–style island of robotic reptiles, glow-in-the-dark sand beaches, sporting arenas where robots battle one another, facial-recognition for tracking every citizen, a cloud seeding system for rain, and even a synthetic second moon that would rise above the city at night.

"I don't need any roads or pavements," the Crown Prince reportedly said in a meeting. "We're going to have flying automobiles in 2030!"[1]

The question is, How much of this is pie-in-the-sky thinking versus realistic planning? To me, this sounds like Las Vegas on steroids, without the gambling and booze. But if all the razzle dazzle can get high-tech companies to relocate to this patch of desert, why not give it a try?

Not to be outdone, China is in the process of building five hundred new smart cities. I've seen some of these cities myself and participated in the planning sessions as a consultant, along with big players like Microsoft, Huawei, Vanke, 3M, Hyundai, and other global corporations. The Chinese government has made it a priority to leverage all the latest technologies to modernize their infrastructure. This is part of a larger strategy to stimulate the economy and make room for rural populations that are continuing to migrate to urban areas for jobs and opportunities.

Xiong'an is China's crown jewel. It's estimated to cost $580 billion to build, topping even the Saudi's oil-rich budget. It's only twenty minutes from Beijing by high-speed rail and will eventually be expanded to cover twelve hundred square miles, which is larger than Greater London. The Chinese government envisions Xiong'an as a new technological hub. This smart city will have an intelligent urban management system. There will also be a state-of-the-art communications network that includes supercomputing and big data. Even more impressive, China's goal is to harness renewable and lower-carbon energy to provide the city with 100 percent clean power.

China and Saudi Arabia aren't alone. Smart cities are going up all over the world. Forest City is Malaysia's answer to Singapore. It is being billed as the city of the future, where there will be no cars and the buildings will be covered in green vegetation. Forest City will accommodate seven hundred thousand residents and span four artificial islands.

Xiang Ye Tao, a manager at Forest City, claims the homes will be so smart they'll keep your orchids perfectly watered without human intervention, and if local children kick a football that breaks your window, it will be fixed before you return home.[2] Whether or not these promises come true, living on a luxury island off the coast of Malaysia has its appeal. But if that's not your cup of tea, there's always Horgos in Kazakhstan, Colombo Port City in Sri Lanka, or Duqm in Oman.

All this sounds great, but how will these cities be any more advanced and efficient than New York, Berlin, Tokyo, or Shenzhen? Is it all window dressing? Or can we now engineer cities in a way that's far superior to anything that exists today? Let's look at some of the innovations in the pipeline.

Most of the new smart cities will probably incorporate smart parking technology—that is, if they have any cars at all. Smart parking solutions range from completely automated parking garages, equipped with mandatory charging stations, to smart parking meters that can let you know in advance when a spot opens up, bill you automatically, and alert you well before the meter expires.

Some cities may opt for electrified pavement, like the one being developed at Stanford University, which can charge not only cars but also fleets of robots that will be zooming around delivering packages. E-pavement will also lower costs for companies like Uber and Lyft, allowing their autonomous electric vehicles to stay on the road for days at a time. It may become so cheap that it won't make sense for most people to own a car.

The real question is, Will you be able to fly to work? Probably not. Using flying cars inside cities will probably remain a novelty for quite a while. They aren't practical for travel within a dense metropolis. Imagine if a flying car has a glitch and slams into an office tower. Most city officials will not want to take the risk of having thousands, or tens of thousands, of flying cars buzzing around. Flying cars will be primarily used to travel between cities, not within them.

In these smart cities, most of the city services will be handled primarily by AI and robots, from managing the electrical grid to maintaining the parks and repairing potholes. The downside of automating everything is that trying to get a human on the phone to respond to your needs may not be so easy. You'll probably have to wind up chatting with an AI if your *Jetsons*-inspired 3D food printer malfunctions.

Sensors will be everywhere in these cities. If the kid down the street is making too much noise pounding on his drum set, an alert may be sent to the appropriate authorities. In Barcelona they are already testing out a system like this to control noise pollution. Smart cameras and sensors will also track every vehicle, rerouting traffic to alleviate congestion and alerting police if there's an accident. If you happen to light up a cigarette in a public space, a smart device will probably tell you to put it out. And don't even think about littering. In Singapore, a city obsessed with cleanliness and order, they already have AI-powered cameras watching for this and automatically doling out fines.

To save energy, indoor lighting, thermostats, streetlights, and practically everything else plugged into the grid will be managed by intelligent algorithms. If no one is on a street or in an office, an AI will automatically adjust the lighting, turn off heaters, close any open windows or doors, and trigger a host of other energy-saving measures.

If there's a hurricane or earthquake, the entire emergency response will be coordinated through a central AI that is fed data from tens of thousands of sensors all across the city. Pollution will also be tracked and managed. Regional factories and other polluters in the area may be automatically shut down during poor air quality days. The waste disposal system will be automated. Cities could use vacuum tubes to enable trash receptacles to whisk away garbage to centralized locations, where garbage trucks can haul it to the dump, a system already being tested in Barcelona.

There will be ecological parks and greenery everywhere. The cities of the future may have more in common with Disneyland than Manhattan. Everything will be monitored, controlled, and orchestrated. This is nice if you crave order, but if you're like me and you appreciate a little chaos and unpredictability in your life, a cookie-cutter smart city might not be your cup of tea. It will be like living in a fishbowl, with all your needs taken care of, except for that of roaming free and perhaps causing a bit of mischief.

SMART GOVERNMENTS

You can't have a truly smart city without smart governance. When I was visiting the Baltic region, I looked forward to meeting with government officials from Estonia because I'd heard so much about the progress they'd made since breaking away from the Soviet Union. I wanted to know how this tiny country of 1.3 million people managed to leapfrog much larger nations and attract business from all over the world.

When Estonia first became independent, they had 1,000 percent inflation. Today, it's close to 1 percent. They also inherited the bloated Soviet-style bureaucracy. Now, they are one of the leanest governments in Europe. They originally had little in the way of private enterprise. Today, they have the most Fortune 500 companies per capita of any country in the world. When inside the Soviet Union, a start-up culture didn't exist. They now have the most unicorns of any other country their size.

What I learned is that Estonia accomplished this by starting from scratch. They reinvented their government from the bottom up. They began by removing price controls, lowering trade barriers, encouraging immigration, and adopting a flat tax. The next step was to bring all government services online. This meant making government transparent. Not only did this reduce corruption and increase efficiency, but it has allowed anyone living inside or outside the country to access government services with the press of a button.

Next, they took the step of implementing an e-Residency permit, so anyone anywhere in the world could easily do business in Estonia without ever setting foot in the country. This was a brilliant move because it enabled businesspeople from all over the world to set up operations in Europe and manage their businesses remotely. Entrepreneurs don't even need a local bank account. They can use any EU bank. Estonia sweetened the pot by allowing foreign business owners to pay no taxes unless they take the profits out of the country. This encouraged reinvestment in the ecosystem.

"We want the government to be as lean as possible," says Ott Velsberg, Estonia's chief data officer. "Some people worry that if we lower the number of civil employees, the quality of service will suffer. But the AI agent will help us."[3]

Velsberg is overseeing the nation's push to use AI to automate government services. Inspectors no longer check on the farmers receiving government subsidies to cut their hay fields each summer. Instead, satellite images are fed into a machine-learning algorithm and overlaid onto maps, then a status update is sent to the farmers via email or text. The entire process is automated.

Another one of the government's machine-learning systems takes the résumés of laid-off workers and matches their skills with potential employers. The algorithms are so good at finding the right matches that over 70 percent of workers who land a job are still employed after six months. This is up from 58 percent when it was performed by humans.

Obsessed with streamlining every process, Estonia automatically enrolls children in local schools at birth. This way parents don't have to sign up on waiting lists or call school administrators.

If that's not enough, Estonia has even rolled out an AI judge to resolve disputes over small legal contracts. No human judges will be involved unless the case is appealed.

"The promise of an AI approach is you get more consistency than we currently have, and maybe an AI-driven system that is more accurate than human decision-making system," says David Engstrom, an expert in digital governance at Stanford University.[4]

Estonia has gone even further, enabling all government databases to connect with each other and share information. The government also allows residents to check who has been accessing their information through a digital portal. This level of automation is paying off. More than two-thirds of Estonian adults file government forms over the internet, which is nearly twice the European average.

Despite bringing all these services online, government officials point out that they haven't suffered a single major data breach or theft. For a small country with limited resources, this is quite a track record. Most big countries, like the United States, aren't even close to this level of government automation.

ROBO COPS: AUTONOMOUS POLICING

Police automation is already happening. PredPol, which is being used in more than forty law enforcement agencies across the United States, claims that its AI can predict where and when specific crimes are most likely to occur in a city, right down to a 500 × 500 square-foot zone.

"We take anywhere from 3–10 years of crime data and run the relevant points of information through our algorithm," stated PredPol. "Long and short-term trends, recurring events, and environmental factors are all taken into account."[5]

PredPol believes its machine-learning algorithms can help police departments make better use of resources and lower crime rates.

"We found that the model was just incredibly accurate at predicting the times and locations where these crimes were likely to occur," says Steve Clark, deputy chief of Santa Cruz Police Department. "At that point, we realized we've got something here."[6]

AI is also influencing how judges dole out sentences. In February 2013 police found that Eric Loomis had been driving a car used in a shooting. After he was arrested, he pleaded guilty to eluding an officer. In determining his sentence, a judge looked not just at his criminal record but also at a score assigned by algorithmic software called COMPAS. Based in part on the AI's prediction that Loomis was likely to commit more crimes, he was sentenced to six years in prison. I'm afraid that Philip K. Dick's *Minority Report* is fast becoming a reality.

The problem is that these systems aren't perfect. Back in 2016, ProPublica reported that COMPAS was racially biased. The software would identify Black defendants as posing a higher risk of recidivism than they do, whereas it would underreport the risk for white defendants. The bias came from the original data fed to the deep-learning algorithms. Knowing how fallible these systems can be, do we really want AI to set bail, determine sentences, and even help pass judgment on the guilt or innocence of the accused? If you're the one being sentenced, probably not. Even if the data turns out to be unbiased, should race be included in the data at all?

It's not just in the courts but on the streets that AI automation is taking an active role. The start-up Knightscope designs and builds fully autonomous security robots. Their fourth-generation model looks like a robotic conehead on wheels. It's packed with sensors and can alert security guards or police if it detects something suspicious going on, but it isn't yet able to apprehend criminals on its own.

"I use the analogy of the police car parked at the corner," says John Santagate, vice president of Robotics at Körber Supply Chain. "Even when no one is in it, people around the car adjust their behavior."[7]

Deterrence may be one of Knightscope's greatest benefits, but it can also gather a lot of information. These robots can scan Wi-Fi signals and identify smartphones and information about their owners. This data can be used to identify and track unwelcome individuals.

Police are also using drones to map out cities, monitor traffic flows, chase suspects, investigate crime scenes from the air, and even perform 3D reconstructions of accidents. Drones can be manually controlled, but increasingly, they'll be autonomous. These flying robots are already helping police with search and rescue operations, emergency disaster response, and delivering relief to victims of hurricanes and earthquakes.

You can imagine the potential of combining ground and air robots to help manage large public events, like a rock concert, monitor protests, and control riots. Police drones are even being used to apprehend other drones that aren't authorized and take them down. And it's not just drones. Ford is developing autonomous police cars. These cars will come packed with sensors and be capable of doing everything from going on high-speed chases to giving out parking tickets.

"Right now, our primary interest is sending the robot into situations where you want to collect information in an environment where it's too dangerous to send a person," says Michael Perry, vice president of Boston Dynamics, which has developed a robotic dog called Spot.[8]

The more sophisticated the AI becomes, the more we'll ask these robots to do. At first, it will be just chasing down criminals and, perhaps, restraining them until officers arrive. However, there may come a point when we enable these robots to take lethal action. Imagine a hostage situation where criminals or terrorists have people trapped. Instead of sending in officers, police may choose to send in robots. Or, what happens if a robot is the first to come across a suicide bomber or mass shooter. Should it just stand by and wait for police to show up, or take action?

In 2016 Dallas police deployed a bomb-disposal robot armed with explosives to kill a sniper who had murdered five people. This was the first time a nonmilitary robot had been used to kill someone.

"We really need some law and some regulation to establish a floor of protection to ensure that these systems can't be misused or abused in the government's hands," says Kade Crockford of the American Civil Liberties Union. "And no, a terms of service agreement is just insufficient."[9]

The more robots we deploy, the more likely these scenarios will come up. Deep automation means there will be less need for human police officers and more intelligent machines taking their place. At some point, we may feel it makes sense to enable these machines to use lethal force, if necessary. The justification will be that the underlying AIs are more accurate and reliable than humans. This isn't hard to imagine. When humans are placed in a stressful situation, fear kicks in, adrenaline starts pumping, and even the best trained officers are capable of making mistakes. Machines wouldn't suffer this biological impairment, and with a well-trained AI, there may be far fewer incidents of unwarranted police violence.

If we know that robots of the future will be less likely to overreact, less prone to prejudice, and less likely than their human counterparts to use force unless absolutely necessary, shouldn't we use them instead? This question will surely be debated around the world, and different countries will make different choices.

What we do know for certain is that robots will be used in increasing numbers to automate and improve law enforcement. And it won't just be robots. The robots will be an extension of an entire system of policing that includes every CCTV camera, microphone, smart device, and computer database in the police network. When an AI makes a decision, it will be taking in data from a huge variety of sources, compiling it, analyzing it, and making a determination of what to do next. A single AI might be coordinating dozens of robots and hundreds of smart devices in the apprehension of a criminal suspect.

Let's hope that most of these future robots, if not all, will be nonlethal. They could be designed to use robotic arms, rubber clamps, nets, and other methods to subdue subjects without causing bodily harm. If we approve lethal machines, they should only be called out in extreme situations, such as if there's a mass shooter or terrorist attack. This way, policing may become more automated and safer. What we don't want to happen is for these robots to fall into the hands of bad actors who use them for ruthless suppression.

If robots are used responsibly, at some point we may come to prefer them over humans. Someday, we may do away with human patrols altogether. This may seem unthinkable, but if you consider a future where we become used to dealing with robots on a daily basis, we may not fear them so much. Instead, we may focus on the statistics. If both the crime rate and death rate plummet in cities that rely on AI and police robots, it only makes sense to use them instead of us.

THE STATE OF SURVEILLANCE: AI IS WATCHING YOU

As we algorithmically automate our world, we won't just have police robots but also surveillance technology playing a central role in our security and commerce. For example, a Japanese start-up has developed AI-equipped cameras that can predict criminal intent. The company claims that stores using its software saw a 77 percent drop in shoplifting losses during its trial run. This seems like a good thing, especially if you're a store owner. The National Federation of Retailers estimates that shoplifting and fraud account for more than $50 billion in losses every year. Clearly, this software is valuable, but when does it cross the line?

Many civil rights advocates are worried about an AI that flags people as suspicious based on statistical correlations, such as how they move about a store, what type of clothing they're wearing, or whether they have tattoos. This can lead to some unpleasant outcomes. How would you feel if every time you entered a store, an AI alerted the staff that you might be a shoplifter based on a statistical correlation? This may cut down on shoplifting, but it can also make certain groups of people feel they are being unfairly targeted.

A Stanford University and MIT study found that many commercial facial-analysis programs demonstrated both gender and skin-type biases. It's extremely difficult to entirely remove bias from AI because much of the bias is embedded in our culture and society. There's even the possibility that the AI itself can perpetuate negative social stereotyping that leads to negative behaviors. Shop employees may become more hostile to certain types of people based on what the AI tells them, and in return, those people may retaliate by committing crimes against those stores.

There is also the possibility that AI may be used to maximize profits at certain customers' expense. Imagine if an AI alerts the staff whenever it identifies a high-value customer entering the store. This could lead to a tiered society, equivalent to a digital caste system, where favoritism and discrimination are practiced on a daily basis thanks to proprietary data hidden away on company servers.

It's not just how you look that may determine your data but what you say. Using video cameras and microphones, AI can monitor your personal conversations with friends as you walk through a store and use this data to build a profile of you. This data may be used to provide better customer service. For example, if the AI detects that you are confused, it may alert the staff to help you. It may also use this data to understand your buying preferences.

At the same time, it's not pleasant to think that everything you say while walking around a store is being recorded and analyzed. Even if no humans are in the loop, the data is still going somewhere. It may wind up being sold to other parties. This happens all the time with online data. There is also no way to know for certain whether this information is completely anonymized. What you say may land in some database somewhere tied to your identity, and there may be nothing you can do about it because you may not even be aware it's out there. Without government regulations in place, it's doubtful we'll have any idea to whom our data is sold or what that third party will do with our information.

There are also concerns around anonymity. Sometimes people want to go into a retail store specifically to buy something with cash because they don't want anyone to know about it. However, if the store is recording everyone with CCTV cameras, this becomes impossible. When shopping, many of the rights and privileges that we have taken for granted are now on the verge of disappearing.

It's not just inside stores that we are being monitored—it's on the road. In New South Wales, Australia, they've integrated machine vision into roadside cameras to spot people talking on their phones while driving. The AI automatically flags drivers, sending out a warning letter to them.

There are now CCTV cameras that can reach out and talk to you by calling your cell phone. These cameras may warn you to keep on the footpath in a park, or they could tell you to stop slacking off at work. There are really no public spaces you can go today where you can be sure you aren't being watched, and privacy concerns are only going to grow more pervasive as deep automation progresses. Imagine a world where robots are everywhere: on the streets, in homes, at schools, and in the workplace. They will be recording everything around them through their sensors, and that data will be valuable. Someone will want to own it, use it, and sell it.

In the UK there is already one surveillance camera for every ten citizens. Estimates vary on the number of CCTV cameras in China, but analysts believe that sometime in the coming decade, China will have one camera for every two people. Is this a harbinger of things to come? Will we reach a point where countries have more cameras in public spaces than people? Imagine what happens when AR goes mass market. At this point millions of people will be walking around with the equivalent of CCTV cameras on their heads, and companies like Apple and Google will be scraping data from everything they see. This will be a privacy advocate's nightmare.

It's not just cameras that we have to think about. Companies like ShotSpotter are using outdoor microphones to pinpoint gunfire location and alert the police. They claim this improves response times and gives officers the data they need. It can also determine how many shooters there are and what class of weapons were fired just from the audio recordings. Their microphones have been deployed in more than a hundred cities across America. The problem is that we don't know where or what they are recording.

In Europe, PASSAnT is deploying smart fences equipped with microphones, which can distinguish between an accidental bump, a stormy day, and someone climbing over.

"Policing will become increasingly tech-based, and right now, not everyone is ready for that change," says Leon Verver, director of the Dutch Institute for Technology, Safety and Security. "But once we can show them this will make their jobs easier and the streets a safer place, I'm positive they'll come around."[10]

If you think it's strange to have microphones throughout a city listening in on your conversations, wait until you hear what researchers at MIT have dreamed up: an AI that can predict what you look like based on your voice. After analyzing a short audio clip of a person's voice, the machine-learning algorithm reconstructs what they might look like. It's not perfect, but it does work.

If that's not creepy enough, MIT has taken it a step further with technology that can see through physical barriers. Using a neural network that analyzes radio signals bouncing off the human body, their system can detect and track a person walking, sitting, or moving behind a wall. This could prove extremely useful for law enforcement and the military, as well as anyone who wants to spy on someone else.

Cortica, an Israeli company, is rolling out its AI-based crime-fighting system across India. Using input from CCTV cameras, it will analyze behavioral anomalies to spot when someone might be about to commit a crime. Their software is based on military screening systems that can identify terrorists by decoding microexpressions: muscle twitches, facial movements, gestures, posture, stride, and other bodily motions.

In China police wear smart glasses with built-in facial recognition. Officers can identify suspects in a crowd by snapping their photo and then matching it to a database. SenseTime has deployed its facial recognition software across China and is now one of the world's most valuable AI companies. With a supercomputing platform comprising more than six thousand high-performance graphics processing units and access to hundreds of millions of faces, it has become adept at identifying people and their actions. SenseTime is now deploying its technology for use in everything from social networking apps to medical services. "Once you have optimized the artificial neural network for facial recognition, it can be run in real-time even on a mobile phone," says cofounder Lin Dahua.[11]

IC Realtime is making it even easier to access surveillance data. Their app runs on Google Cloud and can search footage from almost any CCTV system. Users can download the app, pay a monthly fee starting at just seven dollars, and begin searching video footage.

"Let's say there's a robbery, and you don't really know what happened," says CEO Matt Sailor. "But there was a Jeep Wrangler speeding east afterward. So, we go in, we search for 'Jeep Wrangler,' and there it is."[12]

This type of technology will make massive amounts of visual data easily searchable. In the future, all someone may have to do to find out where you've been and what you're up to is to type in your name. If this technology falls into the wrong hands, it could turn out like an episode of *Black Mirror*. Even if the technology is only used for fighting crime, it will still have an impact on how people act.

But surveillance technology has already found other applications. Hangzhou No. 11 High School in eastern China installed three CCTV cameras to spy on its own students. The system identifies seven different facial expressions: neutral, happy, sad, disappointed, angry, scared, and surprised. The goal is to determine if the students are focused on their lessons. If not, it alerts the teacher.

"Previously when I had classes that I didn't like very much, I would be lazy and maybe take a nap on the desk or flick through other textbooks," says one of the students. "But I don't dare be distracted since the cameras were installed in the classrooms. It's like a pair of mystery eyes are constantly watching me."

This certainly takes some of the fun out of being young. It's like Big Brother has decided no one can ever again throw a spit wad or even nod off for a couple seconds without getting caught.

"It's the same as teachers having an assistant, and it can improve the quality of teaching," says headmaster Ni Ziyuan. "Some have said it can infringe the privacy of students, but it only records students' movements, rather than filming activities in class. And those who focus on lectures will be marked with an A, while students who let their minds wander will be marked with a B."[13]

The same technology might be coming to factory floors, call centers, retail stores, and offices in your city. This may not be a problem for you, but I need time to goof off now and then. It would be incredibly stressful never to have a moment's rest, and I'm concerned that students and workers will begin to suffer from both physical and mental health issues.

Brock Chisholm, a clinical psychologist, has done extensive research on the effects of surveillance on people over extended periods. He has found that the impacts depend on how aware people are that they're being watched, and what they believe the motivations are for the surveillance.

"There's the kind of background, everyday anxiety that builds up, we know it's there, but we kind of ignore it and we don't realize how on edge we were until it's gone," says Chisholm. "For those people, the kind of lower level but building up background anxiety—they're going to have more relationship difficulties, more arguments, they're going to be more hypervigilant, scanning for threats."[14]

Depending on the individual and the type of surveillance, the negative impacts can range from mild anxiety to severe PTSD and depression. Along with these stress-related conditions come a host of physical ailments. Surveillance is not an intrinsically benign technology. If used improperly, it can have a direct impact on our sense of well-being and health.

"We want people to not just be free, but to feel free. And that means that they don't have to worry about how an unknown, unseen audience may be interpreting or misinterpreting their every movement and utterance," says Jay Stanley, an analyst at the American Civil Liberties Union. "The concern is that people will begin to monitor themselves constantly, worrying that everything they do will be misinterpreted and bring down negative consequences on their life."[15]

It won't be long before these deep-learning algorithms are able to determine all sorts of things about our behavior. For example, using CCTV cameras and data from our smart devices, as we walk down a street, an algorithm could predict where we will go, who we will meet, what we may do, and even what our intentions might be. If you couple this technology with ever-shrinking cameras, the smallest now being the size of a grain of sand, all of us will soon be living in *The Truman Show*, a fictional movie where the protagonist has no idea his every move is being captured by hidden cameras as part of a TV show.

Is this the type of world we want to live in? That's a question we need to ask now before what little privacy we have left becomes a thing of the past. The scary part is that it's going to get worse with the mass deployment of tens of billions of smart devices in our offices, homes, streets, and public parks. It will soon be impossible to find any place where we can be sure we aren't being monitored, analyzed, and classified.

That said, there is hope. While we live in a society in which a powerful few can surveil the many, the masses can inversely surveil the few, curbing the potential to abuse this power.

Steve Mann, a professor at Toronto University, who is widely considered the "Father of Wearable Computing," calls this phenomenon *sousveillance*, or undersight. It occurs when ordinary people use technology, like smartphones, AR devices, and wearable cameras, to record the world around them. We see this today, when bystanders use their camera phones to record police violence or when activists use smart devices to estimate the number of people attending a protest and document what takes place. In a world where smart devices are in the hands of everyone, the power of sousveillance can act as a healthy counterbalance to surveillance.

AI FORTUNE-TELLERS: THE PREDICTION MACHINES

Homo sapiens are better at modeling and predicting possible future scenarios than any other species on the planet. But now our intelligent machines are poised to make us look like chimps. AI's ability to consider massive amounts of data far surpasses ours, and that means a computer's ability to gaze into the future and determine what may happen has the potential to disrupt entire industries and reconfigure our lives.

Amazon is using its massive amount of data to algorithmically automate every aspect of its business, including predicting what people will buy. For instance, Amazon can use statistical models to predict how many Gillette razorblades will sell in Chicago next Tuesday, enabling it to ship those products in advance. This is how Amazon can avoid overnight shipping costs and pass those savings along to its customers, who want to receive their orders the very next day. Of course, this works best for high-volume products. The greater the volume, the more accurate these predictions.

Amazon's ultimate mission is to sell its customers products even before they know they need them. How will Amazon accomplish this? It all comes back to the data. Using smart devices like Alexa, as well as data from online and offline shopping, Amazon is seeking to learn and model each individual's buying habits. Once Amazon has enough data, the next step would be for it to begin shipping products to individuals even before they place their orders. Imagine having an Amazon package arrive on your doorstep, and when you open it, inside are the products you were thinking of ordering. If you don't want some of them, you can send them back free of charge.

Amazon will be able to afford to do this as soon as its machine-learning algorithms can become accurate enough to make this profitable. In preparation, Amazon is automating its entire supply chain, from warehouses to the doorstep. The lower the cost of shipping, the more returns it can afford to accept. To accomplish this, it needs to take human labor out of the loop.

Someday, Amazon's warehouses, delivery vehicles, and everything in between will be autonomous. Once this is in place, the cost of customers rejecting an order will be minimal, and Amazon can commence with sending products in advance of ordering. In other words, Amazon wants to take the buying decision out of buying, by automating every step of the process. If its customers stop shopping for products, they aren't searching for the best price online and won't be visiting competing stores. Over the long run, this will prove incredibly profitable for Amazon, while saving customers the time and hassle of shopping.

Google's DeepMind has developed software to predict the likelihood of cardiovascular problems, like heart attacks and strokes, simply from images of the retina, with no blood draws or other tests necessary. It won't be long before there's a wearable device that can tell you to head to the hospital because you're likely to have a heart attack within the next few weeks or to take a certain medicine within the next twenty-four hours to avoid having a stroke.

"Using deep learning algorithms trained on data from 284,335 patients, we were able to predict CV [cardiovascular] risk factors from retinal images with surprisingly high accuracy for patients from two independent data sets," says Lily Peng, a physician-scientist who is leading Google's effort.[16]

DeepMind went on to create an algorithm that can predict if a patient has potentially fatal kidney injuries forty-eight hours before most symptoms can be recognized by doctors. Google's researchers said it already has a 90 percent accuracy. With this type of technology, doctors will stop guessing at what might happen and begin planning for what's coming. Hospitals could also use the data to plan in advance who will be on staff and how many beds they'll need each day of the week, resulting in more lives saved at a lower cost.

It may come as a surprise, but artificial intelligence can even predict your chances of dying within a year by looking at your heart test results. The results may appear normal to a doctor, but the AI can tell if something is wrong. After learning from 1.77 million electrocardiogram results, Geisinger's deep-learning algorithm can now spot abnormalities in the pattern changes that humans tend to overlook and alert the attending physician.

According to the World Health Organization, nearly eight hundred thousand people commit suicide every year. Facebook has developed a machine-learning system to predict possible suicide attempts. Once a post is flagged for potential suicide risk, the information is sent to Facebook's team of content moderators. If the team feels it's urgent, they notify law enforcement so it can intervene before it's too late.

"In the last year, we've helped first responders quickly reach around 3,500 people globally who needed help," wrote Mark Zuckerberg.[17]

Soon, we will see schools and social networks using data to predict everything from the likelihood of students developing eating disorders, like anorexia and bulimia, to whether certain students are likely to commit acts of violence. While there are privacy issues around gathering this type of data and its handling, the benefits are clear.

Facebook is also using prediction engines to maximize its revenue. It ran an experiment asking users what articles, videos, and ads they would click on in the future. Then Facebook asked its machine-learning algorithm to predict the users' future behavior. Who do you think was right more of the time: the users themselves or the AI? The AI, of course.

As disturbing as this sounds, an AI can predict what we will do better than we can ourselves. Despite what we may want to believe, we don't know ourselves all that well. We may think we'll click on the interesting science video, but when the cute kitty video pops up, that's what we wind up watching. The deep-learning algorithms look at what we actually do—not what we say we'll do. Using this data, they can get to know us better than we know ourselves.

What does this mean for our future? It means that we will come to increasingly rely on AI to make decisions in our lives, whether it's about what to watch on Netflix or where to go for dinner. Given enough data, AI will save us from viewing bad movies and eating bad food, and we won't have to spend time reading reviews and trying to sort out what we actually want.

It's a pain for most of us to figure out which product or service is best for our needs. No one wants to hassle, especially if an AI can do a better job. The scary part is that every time we use one of these highly convenient algorithmic services, we will be giving up a little more of our autonomy. It won't matter much at first. After all, if you can find a better movie without having to waste twenty minutes searching for it, what's the harm in that?

But what happens when it's not just a movie now and then, but everything we consume, from groceries and gadgets to news and media? Will machine-learning algorithms wind up determining nearly all of our consumption habits? Is it a good idea to let AI automate our lives? And what does that mean for our ability to make informed decisions?

Remember, these algorithms probably won't be expressly written for malicious ends. They will be designed to please us. But not everything should be pleasing. There are things we need to know that are unpleasant. Also, consumer choice is power. If we stop actively selecting the products and content we consume and hand over this decision-making to algorithms, we will gradually have less and less influence over where our dollars go.

There's also the question of autonomy. At what point do we begin to erode our free will? There may come a time when we begin to delegate all sorts of critical life decisions to our AIs. Imagine if you are fretting over

DEEP AUTOMATION

whether to quit your job. You may be asking, "Is there a better job out there? What can I expect to be paid? If I remain at my current job, what are my chances for career advancement?" Since it knows you better than you know yourself, AI will be able to answer these questions far more accurately than you can. Most people simply wouldn't want to risk making this type of decision without that information. It's just too important to them.

By this logic, the bigger the decisions, the more likely we will be to consult AIs. This could extend into all aspects of our lives, from how we invest our money to whom we wind up dating, or even marrying. Just so you know, there are already AI-driven services out there for investing and dating, and millions of people are using them. As these AIs grow more sophisticated, they will become increasingly invisible.

Relying on AIs will become so ingrained in our behavior that we won't think twice about it. There will come a point where most of us won't even realize it when we're handing over a decision to an AI. In fact, it's already happening. Many apps use AI by default to make decisions for us. If we want to use the app, we have to accept that it will make the choices for us. Already, we're seeing AI-powered apps that program our newsfeeds, share our photos with friends, plan events for us, and prepare responses to incoming messages. There are even apps that can answer your phone calls and speak for you. It won't be long before we can begin delegating our entire day to an AI assistant.

Whether we are trying to decide what to do for the weekend or what to do for the rest of our lives, there will be a machine-learning algorithm with access to vast amounts of data that can help us make a better choice. All we have to do is decide whether or not to use it. And the more we use these algorithms, the more powerful and accurate they will become.

Someday, AI may send an autonomous vehicle to pick up our kids from school and take them to their piano lessons. AI may then make sure our suits are dry-cleaned and groceries ordered. AI may recommend what we should wear on certain days, depending on the events it has scheduled for us. There is almost no area where AI won't be useful and time-saving. The sheer convenience factor will be so high that most people will just accept AI into their lives without a fight, much like we have done with smartphones.

It will be almost impossible to resist. How can you compete, or even function, in a world where everyone else is using AI, while you remain a Luddite on the sidelines? Avoiding AI simply won't be a viable option for most people. Just like businesses must adopt AI-powered automation or else face extinction, we too will either use AI or be left behind. In other words, AI won't just wind up automating our industries but also our lives.

Will we have free will in a world where most of our decisions, both big and small, have been delegated to AI? The answer is yes and no. We will always have a choice, but whether we choose to exercise that right is another question. My belief is that we will choose AI over our own best judgment most of the time. This is because, in the future, AI will become so powerful that its judgment will surpass our own in practically every way.

REINVENTING EDUCATION

Deep automation is transforming education as we know it. In recent years, we've seen how a wave of online education start-ups have made learning cheaper and more accessible. Khan Academy pioneered the way by using online videos to teach math. Now they're teaching everything from physics and chemistry to history.

Duolingo, another educational platform, teaches French, German, English, and more than a dozen other languages for free using a game-like system, where learners can earn points and level up. It has more than three hundred million registered users, making it one of the most popular online learning sites in the world.

Code Academy helps anyone to learn how to program in a variety of computer languages, from JavaScript to Python. With more than forty-five million registered users, it has enabled both kids and adults to learn to code without ever setting foot in a classroom.

IBM's Project Debater is the first AI-powered system designed to teach students how to debate. Its AI can take on a variety of complex topics, ranging from the dangers of climate change to the benefits of universal health care. IBM's goal is to help students build persuasive arguments and make well-informed decisions.

There are also a number of virtual and augmented reality applications designed for use in schools. These apps can take students on tours of ancient Rome, let them experience the age of dinosaurs, or enable them to visualize subatomic particles.

Schools are adopting social software. There's been a proliferation of freemium platforms, like Edmodo and Schoology, that allow teachers to interact with their students and parents, collaborate on projects, share digital materials, and coordinate activities.

China is a huge market for online education. The government has offered generous subsidies, and Chinese venture capitalists have invested more than $1 billion in educational technology start-ups.

Squirrel AI is one of China's unicorns. It's an after-school mentoring platform designed to improve students' test scores. It has opened thousands of learning centers in hundreds of cities. It uses AI algorithms to curate lessons, adapting to students' varying skill levels. For example, middle school math is broken down into more than ten thousand elements, and the AI figures out where students have knowledge gaps and focuses on those.

Due to COVID, a massive shift toward online learning has begun. Teachers are using Zoom and other tools to communicate with their students, but there's still a long way to go. Teachers have yet to fully leverage the latest technologies to transform how students learn. Why is this? It's because very few public schools have the budgets to purchase the latest hardware and software. Schools also tend to be slow to adopt novel technology and teaching methods. This stems from the fact that they are rooted in a bureaucracy that resists change. The system is so heavily weighted toward test scores that school administrators don't want to try anything that might lower their rankings. This is a shame because schools wind up teaching to test instead of teaching to learn. Standardized tests measure only a narrow band of intelligence. They don't take into account a student's imagination, leadership potential, emotional intelligence, oral skills, artistic abilities, or a host of other critical traits that determine a person's true potential.

Standardized classes and curriculums were something that made sense during the Industrial Revolution. They were the only way to scale education to accommodate growing urban populations. Teachers specialized in subjects like math, physics, or history, and gave similar lectures, homework assignments, and exams to all their students, while the students moved from teacher to teacher, much like products on an assembly line.

This was the best we could manage to do a century ago, but with the advent of computers, the internet, and AI, there are far better ways to educate our students. Continuing to compel students to spend eighteen or more years of their lives in classrooms learning how to copy and regurgitate information on exams no longer prepares them for the challenges ahead. Instead, we should be teaching our young people how to create, innovate, and collaborate.

Soft skills will be far more valuable than hard skills in the coming years. It's not facts that matter but how to use those facts. Students need to learn how to think for themselves, lead teams, and overcome obstacles. Understanding how to solve tricky, nuanced problems and navigate the complexities of the real world will be far more important in an automated future than being able to remember the exact dates of historical events or perform mental mathematics.

In economies where technological innovation is accelerating, the ability to quickly identify what's most important, adapt to new environments, and master entirely new systems are what will make someone valuable to future employers and society. We currently do a poor job at teaching this. Most students learn far more valuable lessons on their own and from their peers than they do in the classroom.

If you look at visionary business leaders, almost all of them are lifelong learners and primarily self-taught. Elon Musk read the entire *Encyclopaedia Britannica* at age nine and would pore through science fiction novels for more than ten hours a day. Bill Gates taught himself to code. Larry Page and Sergey Brin attended an alternative school and have attributed their success to its unique focus on discovery and child-centric learning. Most of those shaping our future learned how to creatively problem-solve at a young age and are constantly seeking out new sources of knowledge.

Fortunately, we now have the technology to reinvent education on a massive scale. With tools like machine learning, big data, brain-computer interfaces, and virtual reality, we are on the cusp of being able to turn education on its head by measuring how students actually learn, retain knowledge, and become more creative. In the next decade we have the opportunity not only to improve the quality of learning but to bring down costs, increase student engagement, and provide greater flexibility.

Instead of giving the same courses to all the students, AI-based systems can both automate and personalize the lessons. Unlike teachers, who must design each course to serve a variety of students, AI can give students the choice of what to learn and how to learn it, taking into account each student's abilities and preferences. In most classrooms today, teachers must tailor their courses to the average students, leaving the more advanced students feeling bored and the more challenged students struggling to keep up. AI, on the other hand, can determine exactly what each student has already learned, skip over those parts, and pace the lessons to precisely meet the student's abilities in any given subject.

A well-designed machine-learning platform could also detect a learning disability and accommodate students' special needs, working with them to overcome their limitations. Neural networks are designed to learn through interaction, so over time AI has the potential to craft customized lessons for all types of students and situations, while continually improving the lessons for everyone on the platform. If these algorithms are coupled with biofeedback, brain-computer interfaces, and other biological input, we could begin to understand each student's brain at a much deeper level. The software may even learn how individual students process information, what stimulates their imaginations, and which actions trigger their ability to retain knowledge.

AI-driven systems can also gamify the learning experience. If you look at kids today, they spend countless hours learning how to play sophisticated games and master them. Even social media is a game, where everyone is competing to become popular and liked. This is because games trigger the brain's natural rewards system. Well-designed games know exactly how to motivate students in overcoming obstacles and completing repetitive tasks, like mining for gold and grinding through monsters to accumulate experience points.

The most popular games can get players to invest countless hours and dollars into the system because they understand how to trigger the brain's dopamine centers. In 2010 Henry Chase and Luke Clark released a study showing how dopamine is not linked to pleasure but instead plays a central role in regulating our motivations for pursuing rewards and avoiding consequences. That is why the activities in games don't necessarily have to be pleasurable to be engaging. If we take this mechanism and apply it to education, it has the potential to motivate even poorly performing students to complete assignments.

Deep-learning algorithms also benefit from the network effect. The more students that interact with a machine-learning platform, the smarter the system will become. Imagine being able to gather data from millions of students and figure out what works and what doesn't for all types of people with very different backgrounds, personalities, and intelligence levels. We are just now taking the first steps toward building an AI that can unlock the secrets of how our brains process information and problem-solve.

Once these AI-driven platforms prove that they can teach students better than anything existing schools have to offer, that's when everything will shift. There are already two million students being homeschooled in the United States. This number could skyrocket once these algorithmic educational platforms prove themselves and gain social acceptance.

The tipping point may come when the top universities recognize the data and insights these platforms have to offer. It comes down to college admissions offices accepting AI-generated analytics as proof of a student's potential. When this happens, the bureaucratic walls will come tumbling down, and we can stop relying upon standardized test scores as the gateway to college.

It's not just universities that can instigate this change; it's also the hiring practices of big corporations. If companies begin recruiting new employees not based on what college they've attended but on how students perform in analytics-driven educational programs, it could shift the balance. Most parents will gladly pay to have their kids learn from the best AI teachers and tutors if they believe it can land their sons and daughters a job at Google, Walt Disney, or Goldman Sachs. Why fork out money for private or public universities when an algorithmic education can achieve the same thing faster, better, and cheaper?

No one can know for certain when the first AI-driven platform will break out and change the educational landscape, but when it happens, schools will have to adapt. After this change takes places, students will begin transitioning to online education. Students will still benefit from a structured offline environment for interacting with real people and understanding the world. The educators of the future will be more like mentors, coaching young people as they participate in activities where students work in teams, take on real-life challenges, and learn from one another. These school activities can include everything from group science projects to entrepreneurial programs, theatrical performances, athletic competitions, and leadership workshops.

The power of deep automation to transform how we teach can greatly improve our educational system, and someday rote memorization and standardized tests will appear as anachronistic as riding to work in a horse and buggy.

ALGORITHMIC ART: CAN AI BE CREATIVE?

People have argued that creativity is the last bastion of humanity. Machines may be able to match or exceed humans in other areas, but they can never create meaningful paintings, poetry, and music. After all, how can an inanimate piece of silicon possibly comprehend what it means to be alive? How can an algorithm, no matter how sophisticated, grasp the human experience?

It's a commonly held belief that only humans are capable of expressing the inner truths about our nature and the universe. Our machines may produce something eye-catching, but they can never capture the essence of life because they are not alive. Algorithms are inherently blind to the emotions that color our world, so the art they create will always be lacking.

Although the argument sounds compelling, today's deep-learning algorithms are now challenging this assumption. Researchers at Rutgers University, along with Facebook's AI Research Lab, have developed an intelligent algorithm for creating "paintings." The system analyzed 81,449 paintings from more than a thousand artists from the fifteenth century to modern times. By studying these images, the AI learned to recognize the different styles.

The researchers then developed a creative adversarial network. One algorithm would create new images, and another algorithm would decide if these images qualified as art or merely random patterns. The goal wasn't to create a Botticelli or O'Keeffe. Instead, the AI was programmed to create art where someone looking at the painting wouldn't be able to recognize the artist by name, yet the painting would have all the traits of a master's work. In other words, it would be perceived as an original work of true art.

Once the researchers perfected their algorithms, they recruited both art critics and the general public and asked them to evaluate the images in an online survey. The AI's images were displayed alongside art created by past masters. Without being told which ones were AI-generated, the subjects were asked to rate the complexity, ambiguity, and novelty of the images. They were also asked if the paintings made them feel inspired and elevated, whether the paintings were communicating with them, and if they could see the artist's intentionality. Finally, the researchers instructed the participants to point out the images they believed were AI generated. Surprisingly, many veteran art critics couldn't tell which ones were created by AI much of the time.

"Human subjects thought that the generated images were art made by an artist 75 percent of the time," wrote Ahmed Elgammal, a professor at Rutgers University.[18]

What's even harder to believe is that art collectors are now paying huge sums of money for computer-generated art. Back in October of 2018, an AI-generated canvas print called *Edmond de Belamy*, developed at a Paris-based art-collective, sold for $432,500, more than forty times Christie's initial estimate. Christie's auction house marketed it as the first algorithmically generated portrait to come up for auction. At the same event, an actual Andy Warhol print sold for $75,000, and a bronze work by Roy Lichtenstein sold for $87,500.

Some may discount *Edmond de Belamy* as a publicity stunt. However, if history is any indication, it wouldn't be the first time the naysayers were wrong. In 1917, when the artist Marcel Duchamp introduced *Fountain* to the most sophisticated art critics in Europe, they dismissed it as a practical joke. This wasn't surprising given that *Fountain* was nothing more than a urinal, like those found in any public lavatory. Yet today it's on display at the Tate Modern Museum of Art and considered a seminal work.

Many people wonder why anyone would place a urinal on display in one of the most famous museums in the world. Because it changed how we think about art. It made the point that art is more than handcrafting something of beauty and meaning. It's about transforming how we see and view the world around us. Anything can be art if it speaks to us and opens our eyes to new truths about our culture and ourselves. *Fountain* was produced in a factory, yet because Duchamp chose to present it as art, it is now considered art. Is AI-generated art any different? This is why an art collector was willing to pay a small fortune for *Edmond de Belamy*. Apparently, this collector believes the time for AI-generated art has come.

AI art isn't limited to paintings. Computers are even attempting to write poetry. In 2010 Zackary Scholl, a grad student at Duke University, used a context-free grammar system to compose full-length, auto-generated poems. He then submitted these to poetry websites and noted the reactions of readers. The response was overwhelmingly positive. Scholl went on to submit one of his poems to *The Archive*, a Duke literary journal, and the editors selected it to be featured, not realizing that a computer had generated it. Here's the poem:

A home transformed by the lightning
the balanced alcoves smother
this insatiable earth of a planet, Earth.
They attacked it with mechanical horns
because they love you, love, in fire and wind.
You say, what is the time waiting for in its spring?
I tell you it is waiting for your branch that flows,
because you are a sweet-smelling diamond architecture
that does not know why it grows.[19]

Regardless of what you think of this poem, it's just the first baby step in the direction of algorithmic art. Scholl didn't even use sophisticated AI. Today, teams of researchers can generate poems in the style of almost any writer, including Shakespeare, T. S. Eliot, Yeats, and Wordsworth, by simply feeding their AI enough samples.

It's just a matter of time before our robots are producing sculptures in the same vein as Rodin's *The Thinker* or Michelangelo's *David*. Even dance, theater, and performance art may someday have to share the stage with intelligent robots.

If you find the idea of our machines mass-producing fine art disturbing, you aren't alone. Many people question how a machine, which doesn't even understand what a human being is, can capture and express our innermost fears, desires, and dreams. Won't AI art merely be derivative? And even if some of it turns out to be truly "original," how can a computer match the insights and emotional resonance of a Vincent van Gogh or Pablo Picasso?

The fact is that we have little idea what great artists like van Gogh or Picasso were actually thinking when they created their most famous works. When we go to a gallery and look at a piece of great art, we aren't seeing it as the artist saw it. We are seeing something very different—something based on our own personal experience and the culture in which we are immersed.

Even the person standing next to us will have a significantly different experience of the art, especially if they are from another culture or socioeconomic group. So, does it matter what the artists felt or thought when they created their masterpieces? In truth, most of that information is lost to us. All we are left with is our personal interpretation of the piece combined with what society has told us.

It's easy to dismiss what machines are doing as nothing more than churning out variations on existing themes, but is that a fair criticism? Don't all artists learn by copying and only later branch out on their own? Van Gogh's ideas didn't come out of thin air. He studied Rembrandt and Doré and was heavily influenced by Gauguin and Millet. He wasn't working in a vacuum. Van Gogh's art was a fusion of the ideas and currents surging through his time.

As radical as Picasso's art appears, it also came from the cultural forces swirling about him. He was inspired by impressionism, primitivism, Iberian sculpture, and African tribal masks. He took these forms and began to combine them with his own aesthetics. If you look at the process of creating art, it's always one of manipulating the ideas at hand. What artists do is not that different from how deep-learning algorithms create art. It's not magic. They engage in combinatorial play, blending different styles and ideas to come up with new, fresh approaches.

As our computers become ever more capable of replicating the artistic process, most people will likely come to recognize their work as true art.

It might not be the same as the art humans create, but it won't be any less "artistic."

When people say computers will never create art, they fail to realize that art isn't the artist. Art isn't what you see, hear, or touch. Its physical properties are just the surface. Art is what you feel and think as you experience it. Art is created and resides inside the human brain. We imbue art with emotion. Our brains interpret the images, words, and sounds and transform them into something meaningful. By tapping into society and mining data, algorithms will be able to participate in this dialogue.

Computers have another superpower they can use in the creative process—their ability to understand statistics. While humans are inherently awful at analyzing numerical data in large quantities, algorithms excel at this. To discover what makes art "art," a computer merely needs to turn to us. For example, an AI can post its artwork on a social network, like Facebook or Twitter, and then see how people respond. How long do users engage with it? How many times do they share it? What do the comments say? This type of data can trigger another round of iterations, with the neural network gradually zeroing in on what works and what doesn't.

The advantage of computers over people is that this can happen in real time and at a negligible cost. A powerful AI platform can take in feedback and revise a digital representation of any artform within milliseconds, and then gather more feedback on those changes. No human can redo an entire artwork in a fraction of a second, and then test it out, analyze it, and optimize around specific criteria. In other words, using the power of our computer networks, we have the potential to generate art that is more appealing, compelling, and stimulating than anything we've experienced before. It won't be long before our algorithms can produce art that surpasses anything we can create on our own. This may sound like a radical statement, but it's not. It's merely an extension of the path we are already on.

Using wearable devices, our computers can now record our heart rate, body temperature, perspiration, and other bodily functions. With cameras and other sensors, we can infer a person's emotional state, reactions, and intentions. With brain-computer interfaces, we can record human brainwaves and will even be able to detect specific thoughts and emotions. Our algorithms can use this data to determine exactly how humans will react to any given piece of art and then take that data and iterate on the art until it strikes chords deep within our collective and individual psyches.

Algorithms will even be capable of designing individual pieces of art that tap into our personal hopes, fears, and desires. It all comes down to the data. The more our machines come to know about us, the better they will be at figuring out what we need from art. After all, the data is coming directly from our minds and bodies. We will be telling our machines what art is in the most personal way imaginable. Our machines will merely be doing the manual labor. If we reject a piece of art, the machine will try again and again and again, until it gets our approval. We can argue about what is true art and what isn't, but in the end, art will be whatever we make it. It doesn't matter if it's produced by human intelligence, artificial intelligence, or a combination of both.

My hunch is that fifty years from now, even asking whether computers can create true art will seem absurd. Instead, we will care more about how we interact with the art itself and what role it plays in our lives. The value of art will lie in what it tells us about ourselves and our society, which has been true since our prehistoric ancestors started painting on cave walls. When this happens, the art with the greatest impact and reach, no matter how it is created, will dominate our consciousness.

SILICON VALLEY AND THE FUTURE OF HOLLYWOOD

Does the future of entertainment run through Hollywood Boulevard or Silicon Valley? As it turns out, these two disparate roads are now on-ramps to the same high-tech superhighway. With the rapid rise of online content distribution, we've already seen the entertainment industry undergo fundamental changes. It's no longer enough for the big studios to control the airwaves, cable networks, and movie theaters, the real action is taking place on Netflix, Hulu, Disney+, Amazon, iTunes, and Spotify.

The power of these platforms is raw data and its ability to accurately predict viewing habits. Running a focus group, where you invite a dozen people to watch a television pilot and give their feedback, no longer makes sense. It's far better to gather data on what viewers actually do instead of what they say they'll do. Netflix has proven that overpaid studio execs in cushy Burbank offices can't beat geeky number crunchers in Silicon Valley.

Using data, Netflix has reinvented the studio system. They use deep-learning algorithms to analyze what their viewers are watching and determine the types of shows they'll want next. Armed with these insights, they then head south to Los Angeles and recruit the top talent to produce and star in their content. This merging of technology and human talent have proven to be the winning formula.

Cinelytic, a Los Angeles–based start-up, is helping studios like Warner Brothers and Sony Pictures mimic Silicon Valley by using data to determine which entertainment projects to greenlight. Cinelytic claims that data from more than ninety-five thousand movies and half a million movie professionals enables it to forecast box office receipts within 85 percent accuracy.

An example of this is the reboot of *Hellboy*. Cinelytic predicted it would fail at the box office, and sure enough, it did. Cinelytics deep-learning algorithms estimated that it would gross a meager $23.2 million at the US box office on its $50 million budget. It did even worse, grossing just $21.9 million.

Tobias Queisser, Cinelytic's CEO, worked in the film industry for two years and was continually frustrated by the lack of good data. He felt the industry was using antiquated methods like Excel spreadsheets and endless meetings to sort out the potential of any project. He was used to working in finance, where he had access to real-time data, and saw the opportunity to bring this to film executives.

As Hollywood studios move into streaming in an attempt to compete with Silicon Valley, Cinelytic is giving them a hand. The studios are at a disadvantage because they don't have access to the massive amounts of data that Netflix and Amazon do. To compensate for this, Cinelytic monitors the illegal downloading of content around the world, and they've found a high correlation between what people are downloading and what they'll watch on a streaming channel. They deliver these insights to studio execs, helping them decide which projects will be popular on streaming services.

The start-up ScriptBook is tackling a different piece of the puzzle. It uses machine learning and natural language processing to analyze screenplays, basing its recommendations on four hundred separate criteria, including emotion, protagonist's journey, audience appeal, and act structure. It then determines if the script is worth producing. The start-up claims an accuracy rate of 84 percent, which is three times greater than that of humans. To make its point, Scriptbook ran a retroactive test on screenplays from sixty-two Sony movies. It successfully identified twenty-two out of the thirty-two Sony movies that lost money during that period.

"If Sony had used our system they could have eliminated 22 movies that failed financially," said Nadira Azermai, founder and CEO of ScriptBook.[20]

It only takes ScriptBook around five minutes to analyze a screenplay. It can forecast the MPAA rating, evaluate the characters, detect the protagonists and antagonists, predict the target audience's gender and race, and estimate box office receipts.

"When we show this to customers, the first question is: How is it even possible to give a script to a computer and somehow it can come up with all these outputs?" says Michiel Ruelens, data scientist at Scriptbook.[21]

The secret sauce comes from training the deep-learning algorithm on thousands of past movie scripts and comparing them to the current project. The software isn't perfect; when it retroactively assessed the box office potential of *La La Land*, ScriptBook estimated the film would gross $59 million, when it actually grossed more than $150 million. That said, it did greenlight the film.

Do we really want AI determining what projects get turned into movies? Today's AIs have no idea what drama or comedy is. An algorithm cannot laugh at the jokes or cry during a tragic scene. It only compares one data set with another and has no concept of the real world.

Although the software's financial projections may be accurate in most cases, there is the real possibility of filtering out certain types of content. This is because AIs only look at the past and what was successful. Their data is based on what has already happened, not what could happen. Therefore, it's inherently biased. If it doesn't see a lot of blockbusters with Mexican or Filipino actors, the machine-learning algorithm will just assume that's because they aren't bankable. The danger is that it could become a subtle form of blacklisting.

The same is true for new, experimental ideas. If this technology had been around when Stanley Kubrick, Agnès Varda, Yasujirō Ozu, or Federico Fellini emerged on the scene, they may never have had a chance. Part of the problem lies in the fact that society is not static. Tastes change and trends emerge. What worked in the past isn't necessarily what will resonate in the future. This means ideas that are slightly ahead of their time may be pushed aside for the tried and true.

There's also value in failure. Not every film has to be immediately profitable to be of value to society. *Blade Runner* was a box office flop, costing $28 million to produce but only earning $6 million during its opening weekend. Today, however, it's considered one of the greatest science fiction movies of all time. What a shame it would be if we rely on algorithms to the point where taking a risk on a bold but unproven vision becomes a thing of Hollywood's past.

"Already, we're seeing that we're getting more and more remakes and sequels because that's safe, rather than something that's out of the box," says Tabitha Goldstaub, chair of the UK government's AI Council and cofounder of CognitionX, a market intelligence platform for AI. "A lot of people think it's maths, so it can't be biased, whereas in fact, it's completely the opposite way around."[22]

It's one thing to pick what stories get turned into movies, but what will happen as AI automates the entire production process? Will we still need humans in the loop? Are the days of Hollywood stars coming to an end? Let's start by looking at the music industry. It's much easier to produce an AI-generated album than an entire movie or TV show. A wave of new AI music start-ups has come onto the scene, and they are cranking out AI-generated tunes at a furious pace. These include Beat Blender, Neural Drum Machine, and Piano Genie, which all use algorithms to generate music and beats that sound surprisingly good.

Sony has its own project called Flow Machines. After analyzing tens of thousands of songs, it can compose musical scores and audio stems. The AI was never taught music theory. No one ever told the AI about chords or triads. It learned everything on its own by analyzing existing songs.

"We don't give the machine musical rules or abstract musical knowledge," says Pierre Roy, a senior researcher at Sony. "It's only the machine producing music based on what it learned from the data."[23]

Singer-songwriter and *American Idol* star Taryn Southern was an early adopter of AI music software. When she used Amper to create her song *Break Free*, she began by specifying the instruments she wanted to use, the beats per minute, and the genre. The AI then churned out various musical elements, which she rearranged into a song. Southern also experimented with other platforms, including Google's Magenta, IBM's Watson Beat, and AIVA to produce her first AI-assisted album: *I Am AI*.

"Yes, we are totally cheating," says Southern. "If music is concretely defined as this one process that everyone must adhere to in order to get to some sort of end goal, then, yes, I'm cheating. I am leading the way for all the cheaters."[24]

Is this the future? Maya Ackerman, CEO and cofounder of WaveAI, thinks so. That's why she developed ALYSIA, an app which makes it extremely easy for anyone to create music—regardless of talent. The AI not only generates tunes that fit the users' moods but suggests lyrics, and if users don't feel like singing the song, the AI will do it for them.

Ackerman believes the process of creating music has the power to transform people's lives, enabling them to express pent-up emotions and work through complex feelings. The idea for ALYSIA was born out of her own difficult experiences. As a child, she loved playing piano, but when she was twelve, her parents had to sell the piano and could not afford to replace it. In college she wound up studying computer science but always longed to express herself musically.

"All this time I was sorely aware that something was missing. The artist in me was starving, so I decided to start taking vocal lessons," says Ackerman. "This changed my life. Despite being a fairly awful singer at first, within nine months I became a semi-professional opera singer."[25]

She's now on a mission to help others achieve their musical dreams. As an AI researcher and professor, she came up with the idea for ALYSIA. "As a musician, you really need to have your own music. But even though I worked hard on it for three years, learning to be a producer, taking piano and improvisation lessons, I simply didn't seem to have what it takes to create good original songs," she explains. "Eventually, I realised that my machine-learning skills could make my musical dreams come true."[26]

Ackerman isn't stopping there. Her start-up went on to release *Miss Blue Christmas*, which they claim is the first-ever album where both the lyrics and vocal melodies were created entirely with AI. No humans required.

How good is this AI-generated music? Well, you can be the judge. It is certainly good enough for corporate videos, advertisements, and games, but will you be listening to algorithmic music anytime soon on Pandora, Spotify, YouTube, and other services? Whether you like it or not, you probably already are; you just don't know it. If you look at the biggest costs for online music companies, like Spotify, it's royalties. If they didn't have to pay creators for every song they distribute, their profits would soar. Spotify may have had this in mind when they hired François Pachet, one of the world's foremost experts on the application of AI in the world of popular music.

Pachet was the cocreator of Flow Composer, which evolved into Sony's Flow Machines. In 2012 Pachet also created the first known pop songs composed with AI and launched the first music label dedicated to the professional use of AI for music production. It's only a matter of time before Spotify and others rev up the automated production of music. TikTok took the first step when it purchased Jukebox, another AI-music start-up. If you're a fan of TikTok, you're going to be hearing more and more AI-composed music over those videos.

According to the website Music Business World, Spotify has already been caught deliberately seeding their playlists with music by fake artists. Apparently, Spotify has been paying top music producers to create tracks under obscure pseudonyms. Why would they do this? To get out from under the thumb of the record labels and their expensive revenue sharing deals. Typically, 55 percent of all revenue goes to the music rights holders.

The Spotify-owned tracks each have more than half a million streams, which can add up to a lot of money over time. If a company like Spotify is willing to release human-generated music by fake artists to lower its expenses, there's little doubt it would consider computer-generated music. It may even come up with human-sounding names for the band members, so that listeners won't reject the songs offhand.

There will come a time when these platforms won't need humans in the loop. Leveraging their data, they can easily churn out custom music channels to everyone's liking. If you're a big fan of Taylor Swift, Adele, or Elton John, your personalized radio station could generate endless variations at virtually no cost. It could even blend the user's favorite bands, like the Beatles and Bruno Mars, to come up with unique hits.

The algorithmic music available today is nothing compared to what's coming. Researchers at Osaka University have developed a machine-learning device that analyzes people's brainwaves as they listen to songs. The Brain Music headset examines how people feel listening to various tunes, then adapts to those songs to give them a more emotional and satisfying experience.

Using data from devices like brain-computer interfaces, fitness trackers, and smartwatches, these platforms will even be able to find exactly the right music for your mood, whether it's energized, relaxed, nostalgic, motivated, productive, or passionate. So, whether you're at the gym working out or on the couch relaxing, the AI will have the perfect song just waiting for you to hit play, and it won't cost them a penny in royalties.

This raises an important question: What will hardworking musicians do for a living in a future dominated by AI? They are at a disadvantage because they don't control the means of distribution, and once machines can offer more value than they do, it's going to be tough to negotiate much in the way of royalties. Live shows and concerts may be the last refuge for human performers. People still want to see people on stage, and part of music's appeal is the personalities behind the songs. This will continue to be the case for many years to come, but at some point, when we begin spending more time in virtual worlds with avatars that are as lifelike as any humans we have ever met, even the idea of a live concert will change.

The virtual musicians of the future will likely come complete with an AI-driven personality and backstory. The advantage is that you can spend as much one-on-one time with your beloved artists as you like. They won't be human, but you won't care because they will be as real as anything else in VR. And they will have all the time in the world to entertain you.

We're entering a future where it's not only Uber drivers and factory workers who have to worry about having their jobs automated out of existence; even musicians are at risk. You may argue that the majority of consumers will continue to prefer music created by real people, but that may not be the case. As long as the product is high quality and the price is right, people tend take what the system has to offer. After all, all of us could go and buy handcrafted clothing, furniture, and dishware from artisans, but we usually opt for mass-produced goods. Why should entertainment be any different, especially when it's digital to begin with?

Will movie and video stars be any better off than musicians in the future? Already, we've seen AI's impact on the Hollywood moviemaking machine. If you look at postproduction, AI is helping to automate everything from editing to sound production. And it's hard to find a film without some sort of special effects. *Rogue One: A Star Wars Story* already brought back a CGI version of Princess Leia for a key ending scene. So, what's next? Completely AI-generated feature films?

Yes, that's on its way. James Dean has been dead for more than half a century, but this teenage heartthrob is being resurrected through the miracle of computer graphics to star in a contemporary movie. Why not, if the computer doppelganger looks, acts, and talks like James Dean, and it will sell at the box office? This type of virtual actor will only become more commonplace. As the technology advances, it may be impossible to tell whether we are looking at the real person or a computer-generated clone.

We could even see today's biggest stars make a brief appearance on the set of a movie shoot and then bow to let their AI-counterpart take over. This would free them up to appear in another film being produced at the same time or go to a gala event, and no one would be the wiser. Acting is hard work. It might make more sense to license their persona. In the same way Spiderman and the Incredible Hulk have become big franchises, actors and musicians may license out their brands and spend most of their time promoting themselves and engaging with fans. Using AI and CGI to replicate an actor may become much like ghostwriting is today, where famous people can hire someone behind the scenes to craft their bestselling memoir, while they take all the credit.

Louis Savy, the director of Sci-Fi London, told me about one of the first movies written by an AI. It's called *Sunspring*. It is a short science fiction film about people living in a strange future, where nothing is what it seems. It's not Oscar-winning material, but it is interesting. Savy explained that most people didn't even realize it was written by AI. They just thought it was artsy and a bit weird.

Today's AI-written screenplays leave a lot to be desired, but AI-authored news is on the rise. More and more of what we read online will be the product of smart algorithms, from financial updates to weather reports. More than a third of the content published by Bloomberg News uses some form of automated technology, and this number will only grow. Writers are expensive and slow, and financial data needs to be updated quickly. Timing is often more important than the quality of the prose.

It's not only financial news. Robo-reporters have pumped out articles on minor league baseball for the Associated Press, earthquakes for the *Los Angeles Times*, and high school football for the *Washington Post*. *Forbes* has been experimenting with an AI called Bertie to provide reporters with rough drafts and story templates. With the press suffering from shrinking profit margins, this type of deep automation only makes sense.

China has now launched the first AI news anchor. "'He' learns from live broadcasting videos by himself and can read texts as naturally as a professional news anchor," says Xinhua news agency.[27]

As the technology improves, will AI ever be capable of writing successful screenplays or novels? Having worked in Hollywood as a TV development executive and screenwriter, I can tell you that most movies are pretty formulaic. If you've watched enough, it's not hard to predict what happens next. Writing a character-driven story with realistic dialogue is not something AIs will master anytime soon, but AIs can assist writers in doing better research, generating ideas, analyzing structure, and maybe even outlining a viable plot.

In fact, this is another project ScriptBook is working on. They call their screenwriting software DeepStory.

"We envision a next-generation writers' room where whenever they don't know where to head to for the next scene, they would have Deepstory create it," said Azermai. "The engine takes into account everything that you've written, and it will deliver you the next scene, or the next ten pages, or write it to the end."[28] No matter what you think of AI-generated screenplays, algorithms have proven extremely useful in the movie industry's production and postproduction phases, especially when it comes to automating complex processes, like continuity control, editing, archiving footage, color correcting, touchups, localization, and special effects.

Today, it's hard to distinguish a computer-generated scene or sound effect from an actual one. CGI films, like *Toy Story* and *Frozen*, have already proven successful. The next step may be using AI to generate alternate versions of movies in real time. We already see this in computer games that algorithmically generate worlds on the fly. If you ask gamers, they don't care whether humans or an AI created the world. All they want is great gameplay. Why not use similar technologies for movie creation?

A transition to real-time, AI-generated content will allow for an entirely different kind of viewing experience. Imagine sitting down to watch a movie or TV series of tomorrow. You may be able to choose which actors and actresses you want to appear in the story. Maybe you will pick Leonardo DiCaprio, while your spouse prefers Brad Pitt. So each of you will watch your own version on your own device.

You may want more action and violence, while your spouse may prefer character development and romance. Knowing your preferences, the movies of tomorrow could adapt to suit your individual tastes. Netflix and Disney may even tap our biometric data to further enhance the experience. By measuring our pulse, microfluctuations in our body temperature, brainwaves, and other biofeedback, they could adapt the content in real time to match our moods.

This is already happening in games. A company called Red Meat Games used a heart rate monitor to adjust the scariness of its zombie-infested puzzle-based horror game *Bring to Light*.

"Horror games are scary, in different ways, for different people," explains Keith Makse, the founder. "Using a heart rate monitor allows us to find out what triggers will scare individual players, and using our custom AI, we can change the game in real-time to make the game more or less scary as needed."[29]

Deep automation is going to play an increasingly big role in every aspect of entertainment, from its production all the way through to distribution and consumption. This will fundamentally change how we experience content. There may even come a day when our entertainment is being created and modified as we consume it, adapting to our mood and reactions in a real-time feedback loop. If we go back and stream the same movie or song twice, it won't be exactly the same. I wonder what Orson Wells or Ludwig van Beethoven would think of that.

LIGHTS-OUT FACTORIES AND SUPPLY CHAIN AUTOMATION

China is looking to deep automation as its future. As wages have risen, despite its nearly hundred million factory works, it can no longer compete with lower-cost labor in countries like Indonesia and Vietnam. If China wants to remain a manufacturing hub, it knows it must invest in robotics and AI.

"China is facing a lot of pressures—labor costs, energy costs. And that's why we're putting the emphasis on automation, rather than relying on low-cost labor," says Gerry Wong, who runs a Chinese telecom equipment company. "For R&D people, we told them, from now on, if your product is not designed for automation, it will not get manufactured."[30]

Chinese President Xi Jinping has called for a "robot revolution" in manufacturing, and now it's in full swing. In 2020 China's industrial robotics market was estimated at $6 billion, with close to one million industrial robots in operation.

"China is by far the biggest robot market in the world regarding annual sales and regarding the operational stock," says Joe Gemma, president of the International Federation of Robotics. "It is the fastest-growing market worldwide. There has never been such a dynamic rise in such a short period of time in any other market."[31]

Foxconn, which manufactures the iPhone, is an industrial giant, with more than one million employees and over $180 billion in revenue. Terry Gou, the chairman, plans to replace 80 percent of its workers with robots within the next five to ten years. Foxconn already has tens of thousands of proprietary industrial robots, called Foxbots, in its factories, and more are being added every month.

Gou has always been an aggressive adopter of new technology. He began his business in 1974 in Taipei with $7,500 in start-up money and ten elderly workers, making plastic parts for television sets in a rented shed. He received his first big break when Atari asked him to produce its console joysticks. After that, he traveled to the United States in search of new customers. In the early days, he was known for his aggressive sales tactics, often barging uninvited into companies, only to walk away with new orders—despite having security called on him repeatedly.

Gou is just as aggressive when it comes to automating his business. He's famous for saying that Foxconn "has a workforce of over one million worldwide and as human beings are also animals, to manage one million animals gives me a headache."[32]

While Gou's dream of building lights-out factories is ambitious, it's not out of reach, especially for those factories focused on products with longer life cycles. The technology is advancing quickly enough that he may see the end to his animal-induced headaches within his lifetime. To be fair, Gou isn't as heartless as this sounds. He's formed numerous philanthropic foundations and pledged to give away 90 percent of his wealth to charitable causes.

Another larger-than-life proponent of robotics is Richard Liu, CEO of Jingdong (JD.com), one of China's biggest e-commerce companies. He grew up in extreme poverty. During a speech at his childhood middle school, he revealed how he dreamed of eating meat as a boy because his family only had pork a couple times a year. Without the help of his village, he couldn't have afforded to go to college. "They donated a total of 76 eggs and 500 yuan to send me off for the opportunity that changed my life," he says.[33]

When Liu graduated, most of his classmates wanted to go into government or study abroad. But he didn't want to be a bureaucrat and couldn't afford to travel overseas. He also needed money to pay for his grandma's medical care. This eventually led him to start his own business distributing electronics, which grew into Jingdong.

The company has followed in the footsteps of Amazon, taking a lead in business automation. This has made Liu one of China's richest tech tycoons, with an estimated net worth of more than $7 billion. Like Jeff Bezos, he has become obsessed with delivering products to customers in as little time as possible. The company says that 90 percent of goods bought on JD.com are delivered the same or next day.

"Today we have over 70,000 delivery men working on the street. It's high cost, you know. If you can use robotics to deliver a parcel, the cost will be very low," says Liu.[34] "We aim to eventually have more than one million drones to carry out our various operations."[35]

The company has been experimenting with drones to deliver packages in China for some time. Now, it's expanding to Southeast Asia, where drone deliveries can help it reach remote islands and underserved villages.

"We plan to make use of artificial intelligence and robots to create a business model that is almost totally out of human control," says Liu. "I'm not saying that we can do away with all blue-collar workers. Right now, maybe we need 10,000 customers to make even. But in the future, we may only need 1,000 customers to break even or to make a profit due to automation."[36]

The entire supply chain is being upgraded. Linghao, a Chinese supply chain and logistics company that I advised, is committed to using AI to automate as much as possible. Its Zhitong 3000 platform matches truck drivers with enterprises and optimizes the process. One day, everything from trucks, ships, planes, and trains to warehouses will be fully autonomous. The days of humans bringing packages to your door are numbered.

Nuro, a self-driving robot delivery start-up founded by two ex-Google engineers, is already sending groceries to people's homes using autonomous cars.

"A few years ago, we were telling customers that if they place an order today, we can have it ready for them tomorrow afternoon, and that was okay with them," says Yael Cosset, the chief digital officer for Kroger, a supermarket chain. "Today, some of our customers will expect that same order to be available within the hour."[37]

Moreover, the very way we manufacture goods is evolving. A glowing fictional device called a "replicator" in the *Star Trek* series could create products out of thin air. Today, 3D printing is coming close to that elusive goal. Industrial 3D printers can produce parts reliable enough for airplanes, automobiles, buildings, and other products, using a variety of materials, including metals, glass, ceramics, carbon fiber, plastics, and resins.

Merging manufacturing and retail will result in far greater flexibility in adapting to changes in consumer trends, more individualization of products, and reduced inventory, distribution, and shipping costs. Someday, you'll be able to walk into stores, tell them exactly the style, color, and features you want, and they'll print it out for you on the spot.

DR. ROBOTO: SMART HOSPITALS AND HEALTH TECH

Health care is another critical area that's being automated. *Lancet Digital Health* released a study showing that deep-learning models did at least as good a job as health-care professionals at making diagnoses, while Google AI's system for spotting breast cancer in X-ray images significantly exceeds the speed and accuracy of human radiologists. In US hospitals, this AI system produced a 5.7 percent reduction in false positives and a 9.4 percent decrease in false negatives. This will become true of most AI diagnostic techniques. Humans will be reserved for the tricky cases, but eventually, even those will be better handled by machines.

Using laser imaging and deep-learning algorithms, researchers are now able to diagnose brain tumors in under 150 seconds, as opposed to traditional methods that require a lengthy manual process.

Researchers at the Universities of Surrey, Warwick, and Florence have even developed an AI that can detect heart failure from a single heartbeat with 100 percent accuracy.

At MIT they've developed robots you can swallow. These ingestible origami robots can fold up to fit inside a pill. After they unfold themselves, doctors can steer the robots to clear obstructions, patch wounds, or just look around. Daniela Rus, the first woman to be named director of the Computer Science and Artificial Intelligence Laboratory at MIT, is one of the brains behind this project. A robot evangelist and a natural optimist, Rus sees a future in which robots create robots and even reassemble themselves—changing into whatever shape is optimal for performing the task at hand. She believes that we will soon create smaller, more sophisticated robots that can transform themselves into a slinky that can climb stairs, then change shape, becoming a snakelike robot that can slither through tight spaces, and then convert into a fishlike machine in order to swim.

"What if I told you that we will be able to use robots to perform surgery with no incision, no risk of infection, and no pain?" says Rus.[38]

Rus is working to reduce the cost of robot manufacturing by developing origami robots that are printed in flat sheets and then folded themselves into various shapes in order to perform specific tasks. Her team has been developing lightweight robotic muscles made of tiny airbags that can inflate and deflate. These mechanical muscles can already lift three times their own weight. Imagine attaching them to a skeletal system with joints. These robotic exoskeletons could then enable the robot to climb through a rugged landscape, carry objects efficiently, or perform some highly dexterous operation.

If you visit Rus, you may notice that outside her office sits a small piano, where the students come to play music throughout the day. Originally, it was delivered to her office by mistake. It was supposed to go to her kids at home, but the students loved it so much she decided to make it a permanent fixture. She believes it's important for her students to take a break and get in touch with the right side of their brains.

With a seemingly unquenchable passion for technology in all its forms, Rus can't wait to scale up these origami robots so that they can perform tasks on a larger scale. "The possibilities are endless. And a world with a lot of robots is a world with a lot of fun."[39]

The Dutch start-up Preceyes is headed down a similar path. They've already developed a robotic system capable of performing operations no human can. Using a microrobot, they carried out the world's first surgery inside a patient's eye. This tiny robot acted like a mechanical hand, with seven computer-controlled motors resulting in movements as precise as a thousandth of a millimeter. It will not only allow patients to receive higher-quality treatment but because of its precision, eye surgeons may develop entirely new treatments that weren't possible before the technology existed. It also helps automate the process, which means that eye doctors, especially those in developing countries, can perform a greater number of surgeries at a lower cost.

The next wave of robotic surgeons will be completely autonomous. A team at the Children's National Health System has taken a step in that direction. They've developed a lightweight robotic arm to place a line of sutures all by itself when stitching together intestines during laparoscopic keyhole surgery. The procedure typically requires intricate hand movements that even the most skilled surgeons struggle with. If they miss one, there's an intestinal leak and a serious risk of sepsis. The advantage of a robot is that it never suffers from overwork, a bad night's sleep, a head cold, or distractions in the operating room.

"A surgeon can go click, click, click, these are the places I want a suture to happen," says Animesh Garg, who has worked on surgical automation for the past decade at the University of Toronto in Canada. "We wanted this to be like cruise control of surgery."[40]

As the technology is perfected, this will save money and lives. According to Johns Hopkins, more than 250,000 people die each year from medical errors in the United States alone, ranking third among the leading causes of death, after heart disease and cancer. Another benefit is that many communities, especially rural ones, lack skilled surgeons, and developing low-cost robots has the potential to bring high-quality health care to parts of the world where it doesn't exist. We're still in the early days of robotic surgery, but eventually robots will be capable of doing the bulk of routine surgeries in hospitals and clinics around the world. To give you an idea of the pace of progress, in 2012 robots performed just 1.8 percent of all general surgeries, while in 2018 the number had risen to 15.1 percent, and the pace continues to accelerate. Vinod Khosla, a well-known Silicon Valley investor, believes that robots will replace doctors by 2035.

Even something as intimate and human as mental health is being automated. At the University of Southern California, researchers developed Ellie, a 3D avatar that acts as a virtual therapist. Using machine vision, she interprets patients' verbal and facial cues and responds supportively.

Ellie not only can get patients to open up; she can also detect minute facial and body movements. Distressed people's smiles are shorter and less intense. They also tend to avoid eye contact. Anxious people fidget with their hands. All this data is gathered and analyzed and summarized in a report sent to a human psychologist, who makes the final diagnosis and oversees treatment.

Ellie is the first step toward fully automated mental health care. Eventually, we'll see more advanced versions of Ellie using machine learning not only to gather subtle information, like microexpressions, that a human may miss but also to diagnose the underlying condition and even prescribe drugs, like antidepressants.

This means we could have access to virtual health-care professionals whenever we need them. Just like fitness tracking devices today monitor our exercise routines, heartbeat, and sleep, someday we may have an AI doctor, psychiatrist, and life coach in our pockets at all times. With wearables that track everything from our physical condition to our mental health, advanced machine-learning algorithms will be capable of taking into account the massive amounts of data that our bodies generate and coming up with actionable suggestions.

Imagine a future where a wearable AI doctor can detect when we are about to come down with a cold, advise us to skip our workout and eat certain fruits or vegetables, or even prescribe a specific supplement or medication. If we are feeling depressed, an AI psychiatrist could suggest a change in our diet or recommend an antidepressant. There may be little need for routine visits to hospitals, clinics, and therapy sessions. In fact, we may not go to see a professional human unless it's absolutely necessary, saving billions in health-care costs and countless hours in waiting rooms. Wouldn't that be nice?

ALGORITHMIC AGRICULTURE

Agriculture is another area where deep automation has taken hold. Goldman Sachs estimates precision farming, the combination of agriculture and technology, could be a $240 billion market by 2050. This is a good thing because the World Resources Institute estimates we will need to double food production by 2050 in order to feed nearly ten billion people.

Companies like Descartes Labs and FarmLogs are using big data, machine-learning algorithms, and computer vision to provide farmers with information to help increase crop yields. This technology can reduce the amount of pesticides used by detecting plant diseases before they spread and sending out autonomous drones or robots to spray those specific areas.

According to the California Farm Bureau Federation, farmers using automated tomato harvesters have seen a 90 percent decline in labor costs. And it's not just labor costs. Finding enough workers for the job is its own challenge. In countries like the United States, most people don't want to do the backbreaking labor required to pick fruits and vegetables, so farmers have to rely on migrant labor. Crops have been rotting on vines and bushes because there aren't enough staff to maintain and pick them. This is driving the wave of mechanization that is taking place.

"Whether it's berries or lettuce or grapes, we're all scrambling for labor availability," says Scott Komar, senior vice president at Driscoll's, a berry producer.[41]

Picking berries is one of the most difficult and labor-intensive farm jobs. Berries are delicate, and machines can easily bruise them. It's also difficult to tell which berries are ripe enough to pick. Despite this, Harvest CROO Robotics has come up with a solution using advanced AI and robots. A single Harvest CROO machine is able to pick a plant in just eight seconds. It can cover eight acres in a single day, replacing a team of more than thirty human pickers.

"We'll be picking 24 hours a day, with more picking at night when the berries are cooler and less likely to bruise," says Gary Wishnatzki, who cofounded CROO Robotics with Bob Pitzer, the technological brains behind the machines.[42]

Wishnatzki has worked in the berry industry for nearly five decades. He's a third-generation berry man who sports a white goatee and speaks in a gentle voice with a Southern drawl. Entrepreneurship runs in his family. His grandfather was a penniless Russian immigrant who began peddling fruits and vegetables from a pushcart in New York City. He later established a wholesale business, which included berries.

This gave the young Wishnatzki the idea of entering the berry business by starting his own farm. When he began, a box of strawberries sold for four times as much as it does today. "Berries in winter were a luxury item back then," says Wishnatzki. "And that's where we're headed again unless we can solve our labor problems. I testified before Congress before last year's Farm Bill, and I told them, 'If we don't solve this with automation, we're in huge trouble.'"[43]

That's where the idea for the Harvest CROO machine came in. Wishnatzki believes that robots can make up for the labor shortage by picking strawberries and other crops at a much faster rate than humans. And like his grandfather, he's not afraid to venture into new territory.

The wine industry has already benefited enormously from automation. In 2018 machines accounted for 80 percent of the wine grapes harvested in California, according to the California Association of Wine Grape Growers. Not only does this reduce the farmer's reliance upon humans, but the cost of mechanically picking grapes is less than half that of hand-harvesting. Researchers at the University of California, Davis, have pushed it even further, developing a "no-touch" vineyard. Machines do practically everything, from irrigation to harvesting, reducing labor costs from $1 to only 7 cents per vine.

IoT is also playing a big role in automating agriculture. Increasingly farms are relying on big data to analyze the soil, water, sunlight, weather patterns, plant growth, molds, pests, and so on. The more data they gather, the more efficient they can become. "I think that this is the next great wave of agricultural productivity," says Lawrence De Maria, an analyst at William Blair & Company. "The implementation of precision agriculture with automation will drive yields and reduce input costs for growers. It could rival the Green Revolution and mechanization as great drivers of agricultural productivity."[44]

It's not just fruit and vegetable farmers that are benefiting from technology. It's also ranchers. Sheep and cattle farms in the Australian outback can be enormous. For instance, Suplejack Downs extends across 4,000 square kilometers and is a 13-hour drive from the nearest major town. That's why they're using robots to monitor the herds.

Dairy farms are increasingly using robots to milk the cows. Robots can cut the number of workers on a dairy farm by up to 50 percent. Dairy farming robotics is already a $1.6 billion industry and growing fast.

And it's not just robots. It's also data. Dutch innovation company Connecterra has developed a smart cow collar that tracks a cow's every move. It's like a bovine Fitbit. The system can tell when a cow gets sick one to two days before any visual symptoms arise. It also monitors the cow's behavior and can send out alerts to farmers if the animal does not go for a second helping of hay.

In an effort to make their livestock feel more relaxed, Russian dairy farmers have strapped special VR headsets onto cows. They want to see if this improves the cows' moods and milk production. It's an out-of-the-box idea, but whether it's an effective way to increase the milk production remains to be seen.

Meanwhile, in Rotterdam harbor, the Floating Farm start-up has launched the first fully automated floating dairy farm, complete with forty cows. The company believes that bringing agricultural production closer to big cities is important. "Being in the port, we came up with this idea to build a farm on water," says CEO Peter van Wingerden, "so we could bring healthy food close to consumers, who are increasingly moving to cities."[45]

Van Wingerden came up with the idea when he was in New York working on a floating housing project on the Hudson River. When Hurricane Sandy struck, it flooded New York's streets and crippled its transportation networks. He saw firsthand how deliveries struggled to get through. Within two days, it was nearly impossible to find fresh produce in shops. "Seeing the devastation caused by Hurricane Sandy, I was struck by the need for food to be produced as near as possible to consumers," says Van Wingerden.[46]

Floating Farm's bold thinking doesn't stop there. The whole site is decked out with the latest in dairy tech. This includes self-serve cleaning stations, an automated feeding system, manure-scooping robots, and a smartphone app to monitor the cows remotely.

Van Wingerden is also fixated on self-sustainability. The farm incorporates a machine that separates dry manure material from urine, with the processed dry part used as bedding for the cows and the urine turned into organic fertilizer. The roof collects rainwater, and an array of floating solar panels produces 40 percent of the farm's energy. If that's not enough, the cows are fed a mixture of grass cut from local parks and golf courses, leftover grain from brewers, and discarded potato peelings from restaurants. All of this is automatically cut, mixed, and transported to food troughs by conveyor belts.

Floating Farm isn't alone in wanting to automate sustainable agricultural production while bringing it closer to cities. AeroFarms is one of the leaders in indoor farming. David Rosenberg, the cofounder and CEO, chose to focus on leafy greens because of high rates of spoilage. Most people don't realize it, but 50 percent of all leafy greens grown on traditional farms never make it onto someone's plate. They also have high rates of contamination. Farm-grown leafy greens account for 11 percent of food contamination, which includes listeria, salmonella, and *E. coli*. AeroFarms uses an aeroponic system that can grow greens without any sun or soil, eliminating the chance of contamination and use of harmful pesticides. Instead of soil, the greens are grown on a mesh made from recycled materials, including plastic bottles, and the roots are misted with a solution of nutrients.

Their system uses just 1 percent of the land of a traditional farm and 95 percent less water. This is why AeroFarms has launched its prototype in Jeddah, Saudi Arabia, with the support of Sheikh Saleh Boqshan, and is building an 8,200-square-meter vertical farming research and development center in Abu Dhabi with funds from the United Arabs Emirates. In regions where water is a scarce resource, AeroFarms technology is needed. If they can scale it up and bring down the cost, the benefits to a world that's undergoing climate change are clear.

The start-up Iron Ox is addressing the same problem with different technology. Their 8,000-square-foot indoor facility can produce leafy greens at a rate of roughly twenty-six thousand heads per year, five times that of an outdoor farm. But this is just the first step. Brandon Alexander, the cofounder and CEO, is determined to automate everything, from planting seeds to harvesting.

"At Iron Ox, we've designed our entire grow process with a robotics-first approach," he says. "That means not just adding a robot to an existing process, but engineering everything . . . around our robots."[47]

Alexander is an articulate speaker with a classic Silicon Valley vibe; he wears a sports jacket without a tie and has two days' stubble. Although he doesn't look like a farmer, he has been immersed in agriculture since he was a boy. "Growing up on my granddad's farm—he grows cotton, peanuts, and potatoes in Texas—I often heard that technologies like genetically modified crops were required to scale food production," writes Alexander. "My granddad believed that organic practices do not scale and will not feed the world at an affordable price point. Given the state of technology then, I believe he was right."[48]

It's almost certain that the future of agriculture will be closer to what Iron Ox is pioneering than the farm Alexander grew up on. Humans might not set foot on farms except to fix problems, like malfunctioning machines. Ranchers may take to implanting chips in the brains of their livestock and let the AI control everything. The entire farm may come to resemble a giant high-tech assembly line. Autonomous tractors, seeders, weeding machines, harvesters, and drones will do all the work, while advanced algorithms run the show.

This will cost a small fortune, and many family farms won't have the capital to upgrade to the latest technologies. That means there will come a day when they can no longer compete. Large corporations will increasingly consolidate agricultural production, benefiting from economies of scale. All this will mean greater yields and lower costs for consumers, but we have to ask ourselves, are we losing something in the process? Or is this type of deep automation just the natural evolution of an advanced civilization with billions of people to feed?

Whatever your beliefs, we may look back on the days of family-run farms as something from a more romantic past, like chivalric knights defending their castles or hunter-gatherers living off the land.

ROBOTS TO THE RESCUE

Japan desperately needs robots. Its population is both aging and shrinking. In 2019 births fell by 5.9 percent to 864,000, while deaths rose to 1.376 million. By 2030 one in every three people will be sixty-five or older. As the birthrate falls and life expectancy goes up, the problem will only multiply. There simply aren't enough young people in Japan to take care of the elderly, let alone grow its economy. With a projected shortfall of 380,000 elderly care workers by 2025, the problem is critical. This is why the Japanese government and industry are going all in on robotics.

The Japanese robotics company Cyberdyne has developed a robotic exoskeleton named HAL that helps caregivers lift residents of nursing homes out of their beds. This is critical because one of the leading forms of injury among these caregivers is back strain. The same technology can also be applied to helping elderly and disabled people become more mobile and avoid the risk of falling. As exoskeletons become lighter and more flexible, they may be more common than wheelchairs in old-age homes.

Founder Yoshiyuki Sankai isn't your average billionaire. He sports a mass of wavy black hair and oversized glasses, making him look hip in a geeky sort of way—call him a geekster. He has been obsessed with science fiction since childhood. His love of technology goes back to when he was nine years old and devoured books like Isaac Asimov's *I, Robot* and Japanese manga like *Cyborg 009*. As a boy, he built vacuum tubes and transistor radios, applied electrical stimulus to frog legs, chilled goldfish in dry ice and then tried to revive them, made his own rocket fuel, and attempted to construct a ruby laser. When he couldn't find a ruby among his mother's jewelry, he resorted to melting aluminum oxide, the main material found in rubies, over a burner. The experiment failed, but he never stopped experimenting.

"Since my childhood, I've been thinking about how technology can be applied to people and society to help them," says Sankai, who has spent his entire life working toward the goal of building robots that transform people's lives. "One reason why I can keep going is simple. If this had been a project handed to me by someone else, I'm not sure I could have done it, but this is something that comes from within me."[49]

Despite the menacing name of his company, which happens to be the same as the evil corporation in *The Terminator* films, helping humanity is at the core of Cyberdyne's philosophy. Sankai maintains tight control over every aspect of his business to ensure that its technology is used for peaceful, nonlethal purposes only. Even if it's highly profitable, he won't allow the military or anyone else to use his technology to turn people into robotic fighting machines. Instead, he sees a future where his robots can give the elderly the freedom to live on their own much longer, accompany Alzheimer's patients and help them find their way back to their rooms, enable paraplegics to walk again, and assist workers lifting heavy objects on construction sites, assembly lines, and other jobs.

The timing seems right. The Japanese Ministry of Economy expects the market for elderly care robots alone to reach $3.8 billion by 2035. And this is only the tip of an enormous iceberg. Another area where robots will become indispensable is in helping children to learn. There's already a shortage of qualified teachers in many countries, including the United States. In Singapore some kindergarten teachers now have robotic aides.

"We imagine a future not too far off, where interactive robots with the ability to perform multiple human tasks and provide visualization of complex ideas can help children to learn and collaborate better," says Foo Hui Hui, a government official in Singapore.[50]

Many of these robots are designed to engage children in a way that smartphone apps and computer programs cannot. Kids can hold them, take them apart, and put them back together in different ways. These smart toys allow children to learn by manipulating and interacting with physical objects, which engages a different part of the brain. Bee-Bot, a black-and-yellow-striped robot, can teach kids Mandarin, while Rubi, a bug-eyed boxy robot with a touchscreen, can perform songs and play games, helping children to improve their vocabulary. Rubi can even display emotions. If a child yanks off its arm, it's programmed to cry out as if it has been hurt.

In the United Kingdom there is a robot called Nao that helps autistic children, who tend to struggle in social settings, learn how to pick up social cues and interact with others. Unlike most people, robots have endless patience and can accommodate the special needs of children with disabilities.

"Even the most well-intentioned and kind teacher, mother, or father has a limit," says Solace Shen, a psychologist who studied robot-human interactions at Cornell University. "Once you repeat something for the 10-millionth time, you're not going to want to do it again. But robots don't have that problem, and children love the repetition."[51]

Kids who are too ill to attend school are now able to send a robot in their place. At a school for children with special needs in Copenhagen, Denmark, they're using telepresence robots to allow children stuck at home or in a hospital to interact with their peers and teachers.

It's not just in schools and nursing homes but also on construction sites that robots are being employed. Japan is suffering a shortage not only of caregivers but also of construction workers. Even in the United States, construction companies are struggling to find qualified laborers. In 2019 around seven million US construction positions went unfilled.

Fortunately, robots are coming to the rescue. There are now robotic welders, carriers, and all-purpose lifting tools being used on construction sites. There are even robots that monitor jobsites in real time to determine how much progress is being made. Hadrian X, a bricklaying robot, can build all the walls for a house in just three days. SAM, another bricklaying robot, can lay four times as many bricks as a person working alone. And TyBot can continuously tie steel rebar all day long without getting tired, freeing up humans from this backbreaking labor.

Construction robots tend to be large, but robots used in other industries come in all shapes and sizes. Rolls-Royce, which makes jet engines, is developing tiny robots to keep aircraft flying longer. Equipped with cameras and 3D scanners, these roach-like robots will be able to climb inside jet engines, scope out problems, and remove debris.

"They could go off scuttling around reaching all different parts of the combustion chamber," says James Kell, a technology specialist at Rolls-Royce. "If we did it conventionally, it would take us five hours."[52]

Nothing seems to be out of the reach of AI and robotics. Eijiro Miyako once dreamed of becoming a filmmaker like Steven Spielberg when he was in high school. Putting his imagination to good use, he works at the Japan Advanced Institute of Science and Technology, creating tiny flying robots. One of the insect-sized drones that he's designed is capable of artificial pollination. Coated with a patch of horsehair bristles and an ionic liquid gel, this tiny flying robot can collect and transfer pollen from one plant to another. Miyako hopes a fleet of these may eventually help farmers survive the collapse of honeybee colonies around the world.

The word robot may bring to mind an image of a mechanical object, mostly made of metal or hard plastic, but not all robots have to be made of rigid materials. Researchers are working on building soft robots. These can bend, twist, and even stick together. Like squishy Lego blocks, they can combine in various ways forming new, more sophisticated machines.

"There's a lot of interest in materials that can change their shapes and automatically adapt to different environments," says Ian Wong, a professor at Brown University. "So here we demonstrate a material that can flex and reconfigure itself in response to an external stimulus."[53]

There are even robots made from biological cells. A team of researchers have developed the world's first living, programmable organisms. Called xenobots, they can move, pick up micro-objects, and even heal themselves after being cut. A supercomputer running an evolutionary algorithm designed these millimeter-length robots. Living heart muscle cells expand and contract on their own, powering the robots for a week. One day scientists hope xenobots will clean up microplastics in the ocean or even deliver drugs inside the human body.

Robots are even beginning to learn from us. At the University of California, Berkeley, researchers have developed a robot that can quickly learn to copy humans. Using a machine-learning algorithm, the robot can watch a single video of a human picking up an object and then mimic the actions precisely. This technology makes training robots to do various on-the-job tasks much simpler.

"By applying meta-learning to robotics, we hope to enable robots to be generalists like humans, rather than mastering only one skill," explains the team at University of California, Berkeley. "Meta-learning is particularly important for robotics, since we want robots to operate in a diverse range of environments in the real world."[54]

At MIT, they've built a system whereby a human pilot wearing an actuator suit can "drive" a humanoid robot. A pair of VR goggles and a camera mounted on the robot's head lets the pilot see whatever the robot sees. As the driver moves his or her arms and legs, the robot copies every action, gathering more data. Once the robot gathers enough data, its machine-learning algorithm will be able to perform the same tasks as the human on its own.

Scientists at Nanyang Technological University in Singapore are building robots that can learn just by reading an instruction manual or looking at an image of an assembled product. The ultimate goal is to make adaptive robots that can learn in much the same way humans do. This way, having a robot on the job will be like hiring a new employee. With some training, the robot will be able to do pretty much anything that needs to be done.

E-commerce giant Alibaba has built a hotel almost entirely staffed by robots, which can do everything from serving food and drinks to providing room service. It's called FlyZoo and is located in the city of Hangzhou, China. There is almost no human contact, as guests make reservations from a mobile app and enter rooms using facial-recognition technology. If you don't mind forking out a couple hundred dollars a night, you can stay there and see for yourself.

If that's not enough, the age of robot restaurants may be coming. Spyce is a perfect example. Three MIT grads founded the start-up in the basement of their fraternity. There they constructed a Rube Goldberg–like prototype that used microcontrollers, inexpensive oven hoods, plastic trash bins, power strips, and a household air conditioner to cook up meals on demand. Today, they have a high-tech robotic restaurant in Boston that serves up tasty vegetarian, pescatarian, and gluten-free meal bowls.

Like all college students, the Spyce Boys, as they're called, didn't have a lot of money to spend on food. They were tired of spending fifteen bucks on meals that didn't taste that good and weren't always healthy. What they wanted was low-cost, super tasty health food, and that's exactly what Spyce has delivered. They offer seven different bowls for $7.50 each. They can keep the cost low because the cooking is almost entirely automated. A gleaming robotic contraption takes care of everything: it measures the right portions of each ingredient, dispenses them into woks, stir-fries them, and then drops the finished meal into a disposable serving bowl. Voilà! It's ready to eat.

In addition to attracting a large amount of venture capital, the Spyce Boys have brought celebrity French chef Daniel Boulud on board as their culinary director. He's known for his Café Boulud in New York City, which has earned a Michelin star. Spyce's menu includes zesty Latin, Thai, Indian, Moroccan, and Lebanese bowls, and it's now planning to expand across the country.

Could this be the future of fast-casual restaurants? With the spread of COVID-19 having caused so much disruption to the food industry, this trend of having robots prepare meals will surely accelerate. Already, there are dozens of start-ups rethinking how we eat. Blendid's robot cranks out custom smoothies. Wilkinson Baking's BreadBot does the work of a full bakery. And Creator's robotic kitchen makes the perfect burger, including grinding the beef, frying patties, toasting buns, dispensing condiments, and assembling the sandwich.

It's not just meals where robots can serve you. Have a toothache? In China they now have a robotic dentist that can install dental implants.

Suffering from kidney trouble? At the University of California, San Francisco, they've invented the world's first implantable robotic kidney. Need a babysitter? Aeon, a major Japanese retailer, has a robot that can take care of your kids while you shop. Want a robot maid, like in the *Jetsons* cartoon? Aeolus's humanoid robot can not only use a vacuum cleaner to spruce up your home but also grab you a cup of tea.

There's almost nowhere in our society that AI and robots can't step up and pitch in. Not only that—they tend to be more capable and efficient at specific tasks; they aren't lazy; they can work nights and weekends; they don't require overtime pay, sick leave, or vacations; they never complain or talk back; and they only take breaks to recharge. The question is, Will they be taking your job anytime soon?

THE FUTURE OF WORK: A JOBLESS SOCIETY?

If you work as a bank teller, cashier in a supermarket, or forklift operator, you probably already know there's a robot gunning for your job. Automated teller machines, automated checkout machines, and autonomous forklifts are everywhere. Despite this, the majority of Americans aren't afraid that robots are coming to steal their jobs just yet. However, that could change over the next decade, especially as the next generation of robots become more adept at performing both blue- and white-collar work.

McKinsey Global Institute estimates that up to a third of the American workforce will have to switch to new occupations by 2030. That's fifty-four million US jobs that are at risk in the coming decade. China faces an even bigger challenge. According to World Bank research, automation threatens up to 77 percent of Chinese jobs. No one knows for sure what will happen, but what is certain is that with each advance in AI and robotic automation, the numbers will go up.

It no longer matters if you're a factory worker, Uber driver, or medical doctor, eventually a machine will be able to do your job better than you can. McKinsey found that about half of all work activities globally have the potential to be automated in the next decade. Worldwide, 800 million workers could be displaced, and up to 375 million may need to learn new skills. If you are one of the unfortunate ones who cannot adapt, you'll be out of a job.

High-wage workers are expected to be less affected by the sweeping changes because they can adapt their skill sets more easily to the new technologies that emerge. Low-wage earners will also be less affected because they are flexible and are more economical than costly machines for many tasks. It's the middle-income earners who are at most risk of losing their jobs and not being able to find new ones.

We're already seeing this in the United States. Many of the middle-income jobs are disappearing and being replaced by gig, temp, and contract jobs, which offer few, if any, benefits and not much in the way of job security. The income inequality will only get worse in the coming decades as the gap widens between high-paid, high-skill jobs and very low-paid, low-skill jobs.

We may go through recessions, but the authors of the McKinsey study and many other economists believe that things will bounce back, and job growth will continue through 2030. That may be so, but eventually there will come a time when our machines become so efficient and inexpensive that they can outcompete even the lowest-paid workers, while being so capable and versatile that they can take on many high-skilled, highly compensated jobs.

They call this point economic singularity—or the end of work as we know it. It is the natural outcome for a technologically advanced capitalistic system, where profits are the ultimate goal. As we near this point, governments will have to rethink how to restructure our system. If they don't, there will be mass discontent, as billions of people suffer in a state of permanent unemployment.

Andrew Yang, who campaigned to be president of the United States in 2020, believes we should grant every American, regardless of need, a universal basic income of $1,000 per month. But will this be enough for a future with markedly fewer jobs? Governments will need to rethink social programs for the coming age of automation.

When the crisis eventually materializes, corporations will have to fork out enough money for a universal basic income that people can comfortably live on. There's no other way. If governments don't start taxing companies at an appropriate level and redistributing the wealth in an equitable manner, we'll wind up with a handful of extremely wealthy billionaires and technocrats, while the rest of the population lives in dire poverty. That's not a future most of us look forward to.

Eventually, corporations, which always push for lower taxes, will have to capitulate. After all, there will be practically no one who can afford to buy their products and services if they don't participate in wealth redistribution. Ironically, capitalism may end up transforming into a sort of neo-socialism, where the vast majority of people are paid not to work. This wasn't what Karl Marx envisioned for his workers utopia, but it just may be the natural endgame.

Some may argue that this would be a disaster. They worry that a population without work will become indolent, discontent, and depressed. But I don't see this happening. People have a remarkable ability to find meaning in whatever they do. I see a future where work will no longer play a major role in how we determine our self-worth. The reason jobs matter so much to us right now is because our society judges each of us by what we do for a living. In the future, when the vast majority are no longer working, social norms will adapt. We will no longer expect our peers to have jobs, so what people do for a living won't matter.

Instead, we'll live in a world where people define themselves not by their jobs but by whatever they choose to do. For some, it will be their hobbies. They may strive to improve their athletic abilities by competing in marathons or lifting weights. Others may become experts on ancient cultures, study astronomy, or take up dog breeding. I imagine that a large number of people will pursue their religious beliefs, spending time engaged in spiritual pursuits, like rituals, prayers, meditation, and pilgrimages. There will be no shortage of things in this world that can provide meaning, from spending more time with family and friends to traveling the globe.

If you look at most jobs today, they aren't that exciting. People working on an assembly line or flipping burgers at a fast-food joint typically don't love their jobs. They may appreciate the income and enjoy the comradery of their coworkers, but the work itself isn't what gives their lives value. And even those who do find great meaning in their work can find the same thing in other pursuits.

People are always creating meaning for themselves. Fifty years from now, I'm certain that sports will be more popular than ever. Watching robots play basketball, soccer, or tennis isn't very interesting. I also think we'll continue to reward our top athletes with fame and huge perks. The same will be true for other types of entertainers. It's one thing to watch a robot sing or dance, but people will always prefer humans at some level. Live shows of all types will become even more highly valued. This includes everything from concerts and theatrical performances to video game competitions. There will also be entirely new activities that we can't even imagine today. Many people may spend their lives competing in virtual reality to become superstars.

If the transition to a fully automated workforce can be managed gracefully, I see a bright future where each of us is free to pursue whatever interests us most, and everyone has enough income to live a satisfying life. I don't think we should worry about preserving jobs as we understand them today. Instead, we should be planning for how people will exist in a post-capitalist society, where human labor is no longer required, and everyone has the freedom and time to pursue whatever they value most.

FORCE 5

INTELLIGENCE EXPLOSION

Intelligence Explosion, the force driving us to develop new forms of superintelligence that far surpass human capabilities, will bring into existence sentient machines that run our economies, act as our champions, and merge with our consciousness.

In the previous section, we looked at how algorithms are being used to automate human civilization. In this section, we'll take a leap into the future and explore whether our machines will become self-aware. What will our world look like when machine intelligence equals or exceeds that of humans? And how might these super-sentient robots come to view humanity?

The first thing to recognize is that even with all the remarkable things that computers can do today, they aren't close to matching the human brain.

Getting AI to the point where it can equal a human's basic understanding of the world is extremely difficult. For decades, computer scientists have talked about building a superintelligence that can successfully perform any intellectual task that a human can; however, this goal still remains elusive. All the AI breakthroughs we have discussed so far are what we call artificial narrow intelligence (narrow AI or ANI), which is a machine that outperforms humans in some very narrowly defined task. What we see today are countless ANIs working separately and in concert to help us solve problems and automate tasks.

Because narrow AI is so powerful, it often gives us the impression that it's more capable than it actually is. However, this illusion can easily be broken if you pose a question to a narrow AI outside of its specific domain. Try asking an autonomous car what a road is. It won't have a clue. Go ahead and ask Siri or Alexa what a computer is, and it will give you some canned response without truly understanding what it's talking about. The smartest narrow AIs can be compared to all-powerful but painfully stupid wizards. They can perform miraculous feats without any clue as to what they mean. Even three-year-old children can comprehend more about the relationship between themselves and the world than today's top machine-learning algorithms.

Will this ever change? Will there come a time when AIs can match or even exceed our comprehension of the world? Perhaps, but it's not going to happen within the next few years. There are some computer scientists who argue that superintelligence will never come about. They believe we are already at the limits of the current generation of machine learning algorithms and for a computer to truly think and perform like human beings is a fantasy. I'm not in that camp. I think it's only a matter of time before we come up with new algorithms and techniques for enhancing machine intelligence—especially given the amount of resources we are pouring into AI development.

In a study entitled "When Will AI Exceed Human Performance? Evidence from AI Experts," the authors asked 353 AI researchers from around the world at what point will AI be able to do absolutely everything a human worker can do but more efficiently. An average of all the answers came to the year 2060. That's not tomorrow, but it's not that far off.

When discussing superintelligence, keep in mind that it isn't limited to machines that are intellectually capable but also emotionally intelligent. By 2060 will our computers be capable of empathy? If so, how will these empathetic machines change our relationship with technology? Is it possible that humans could fall in love with a humanoid robot—and could that robot love us back?

These are just a few of the questions we'll ask as we take the leap into a future where machines can think and behave like people. By the end of *The Five Forces* we'll have explored everything from giant organoid brains to a runaway superintelligence, as we debate the practical and philosophical implications of creating an artificial consciousness that is not only exponentially smarter than us but possesses a will and desires of its own.

SINGULARITY: THE SUPERINTELLIGENCE IS COMING

Most everyone agrees some form of superintelligence in on the way. The big question is, When? It's one thing for an AI to beat the world's best *Jeopardy*, Go, or chess champion and another for it to understand the world or learn any arbitrary intellectual task as well as human beings can.

"We're still pretending that we're inventing a brain when all we've come up with is a giant mash-up of real brains. We don't yet understand how brains work, so we can't build one," says Jaron Lanier, a pioneering figure in the world of VR.[1]

There are a number of experts who believe this type of superintelligence is still a long way off, perhaps a century or more, while others are convinced it's around the corner. The dates range from as early as 2029 to 2200 and beyond.

"Some people in the deep learning camp are very disparaging of trying to directly engineer something like common sense in an AI," says Martin Ford, author of *Architects of Intelligence*. "They think it's a silly idea. One of them said it was like trying to stick bits of information directly into a brain."[2]

Elon Musk, of course, takes the opposite view: "The biggest issue I see with so-called AI experts is that they think they know more than they do, and they think they are smarter than they actually are. This tends to plague smart people. They define themselves by their intelligence and they don't like the idea that a machine could be way smarter than them, so they discount the idea—which is fundamentally flawed."[3]

Stuart Russell, a professor at University of California, Berkeley, AI thought leader, and author of *Human Compatible*, puts it another way, "I always tell the story of what happened in nuclear physics. The consensus view as expressed by Ernest Rutherford on September 11th, 1933, was that it would never be possible to extract atomic energy from atoms. So, his prediction was 'never,' but what turned out to be the case was that the next morning Leo Szilard read Rutherford's speech, became annoyed by it, and invented a nuclear chain reaction mediated by neutrons! Rutherford's prediction was 'never' and the truth was about 16 hours later. In a similar way, it feels quite futile for me to make a quantitative prediction about when these breakthroughs in AGI [artificial general intelligence] will arrive."[4]

Experts aside, I'm convinced that most babies born today will live to see some form of artificial superintelligence. But how will this come about? There's a consensus that we'll see an intelligence explosion as soon as AI matches or exceeds our understanding of the world and begins to rewrite its own code.

Unlike human beings, where it takes 18 years just to graduate high school and a generation to pass down our genes, computers can share knowledge at an astonishing rate. Once a single machine learns how to do something, it can pass this knowledge along to every other machine in its network. For example, if we send a robot to Mars, and it figures out the best way to move about on Martian soil, instantly, every other robot on the red planet can have this knowledge.

Imagine hundreds of thousands of self-learning robots figuring out how to solve hard problems. As they share information, the feedback loop would further accelerate learning. If you couple this with new manufacturing techniques using advanced 3D printers, companies could potentially crank out improved versions of robots on a daily or even hourly basis. This means the robots of today will look like children's toys compared to what we'll see someday.

At some point, we won't program computers. Human coding is slow and prone to errors. If an AI can code without people, the process will speed up exponentially. And once an AI can learn to code ever more intelligent versions of itself, we'll hit an inflection point called AI singularity. This is where a runaway reaction of self-improvement cycles can lead to an intelligence explosion, resulting in a powerful superintelligence that qualitatively far surpasses all human intelligence.

Hypothetically, the only limit to how smart this AI can become is the raw computing power and storage at its disposal. Hopefully, this hyperintelligent AI will show more compassion for us than we have for our fellow inhabitants of the planet. If so, we will be in good hands. If not, all bets are off.

CAN MACHINES BE CONSCIOUS?

Let's delve into a long-debated question: Can machines ever become conscious? It's one thing for a machine to be capable of solving hard problems; it's another for it to become fully self-aware. Will a machine someday be able to utter Descartes's famous words, "I think, therefore I am," and truly comprehend this in the same way that human beings do?

Let's run a thought experiment. Take your best friends and imagine that tomorrow they come and tell you they are robots. Would this change how you view them? Just by saying this, would you come to view them as machines with no true feelings or self-awareness? Or, after some reflection, would you continue to treat them as conscious beings? And would you still care for them and expect them to feel the same way about you?

Just like in the series *Battlestar Galactica*, someday we may not be able to distinguish robots from humans. Our machines may become so adept at impersonating us that there's no outwardly discernible difference between them and us.

Believing that only biological organisms can possess consciousness may come to be seen as human exceptionalism. We have a long history of putting ourselves at the center of the universe. It's only in the last couple of centuries that we've begun to recognize that there's nothing that special about human beings. We're just slightly more intelligent apes stuck on one of trillions of planets scattered throughout the universe.

Even if we believe robots can never achieve consciousness, does it matter? If their actions are indistinguishable from those of humans, we would naturally interact with them in the same way we do one another. They may even be conscious, but in a very different way than humans are. All we can assume is that their internal processes are distinct from our own, but how and in what manner is unknowable.

Giulio Tononi of the University of Wisconsin has developed what he calls the integrated information theory of consciousness. Under this theory, all physical systems that are sufficiently integrated and differentiated will have a minimal consciousness associated with them. This theory doesn't discriminate between biological cells and other materials, like silicon circuits. In other words, as long as the causal relations among the transistors and memory elements are complex enough, computers should achieve some degree of consciousness.

If this theory turns out to be correct, then consciousness is an emergent property of complex systems and is not exclusive to human beings or even biological organisms. Anything from your thermostat to a potted plant may contain a sliver of consciousness. It wouldn't be anything like human consciousness, but it would still be aware to an extremely limited extent of the environment in which it exists.

"I conjecture that consciousness can be understood as yet another state of matter," wrote Max Tegmark, a Swedish-American physicist, cosmologist, and machine-learning researcher at MIT. "Just as there are many types of liquids, there are many types of consciousness."[5]

Where Tononi's theory diverges from the popular conception of an AI becoming sentient is that it precludes simulations from ever achieving consciousness. The distinction here is that under integrated information theory, consciousness is a fundamental property of the universe that arises from matter. It cannot be simulated in software. Proponents of this theory believe that no matter how intelligent an AI program becomes, it will never obtain a true awareness of itself or its surroundings. It may appear conscious, but it will never be sentient.

Tononi's theory has certain sympathies with panpsychism. The panpsychists believe everything is conscious. This theory has enjoyed a revival because it offers a compelling middle ground between the ideas expressed in dualism and those in physicalism. Panpsychists believe that mentality is fundamental and ubiquitous in the natural world and consciousness exists on a continuum. This means even electrons and quarks contain the tiniest glimmer of awareness and have experiences unique to their nature.

"Computers could behave exactly like you and me," says Tononi, "Indeed, you might be able to have a conversation with them that is as rewarding, or more rewarding, than with you or me, and yet there would literally be nobody there."[6]

Functionalists, on the other hand, believe a digital simulation is enough to bring about consciousness. According to functionalism, mental states are identified by what they do rather than by what they are made of.

"I am a functionalist when it comes to consciousness. As long as we can reproduce the relevant relationships among all the relevant neurons in the brain, I think we will have recreated consciousness," says Christof Koch, president of the Allen Institute for Brain Science. "It's more likely that we have to recreate all the synapses and the wiring [connectome] of the brain in a different medium, like a computer. If we can do all of this at the right level, this software construct would be conscious."[7]

Daniel Dennett, author and codirector of the Center for Cognitive Studies at Tufts University, agrees, "I think that conscious AI is possible because, after all, what are we? We're conscious. We're robots made of robots made of robots. We're actual. In principle, you could make us out of other materials. Some of your best friends in the future could be robots."[8]

Despite having philosophy degrees from both Harvard and Oxford, Dennett is a self-described autodidact. He believes the mind is a collection of computerlike information processes, which happen to take place in carbon-based rather than silicon-based hardware.

The difficult part is that the smartest scientists and philosophers can't even agree on what consciousness is. We all believe we're conscious, but then again, consciousness itself may be an illusion. It might be no more than a by-product of our brain's function—an aftereffect of our neural circuitry. For all we know, everything is determined through the interplay of our biology and environment. We may be just passive observers of our actions, but through some trick of brain chemistry, we feel like we're making choices.

This theory is reinforced by fMRI scans of the brain that show how our brain prepares for a decision several seconds before people feel like they actually made the decision. This deterministic view goes against our intuitive understanding of ourselves and the world, but it doesn't mean it is untrue.

"I'm a robot, and you're a robot, but that doesn't make us any less dignified or wonderful or lovable or responsible for our actions," says Dennett, who argues the self is nothing more than a convenient fiction that allows humans to integrate various neuronal data streams. He contends that our subjective conscious experience of the world, the blueness of blue or the painfulness of pain, are nothing but illusions.[9]

So, will our machines ever be conscious? Whatever you choose to believe, it's inevitable that our machines will eventually come to look and act like fully conscious beings. Just like we give one another the benefit of the doubt, I believe we will come to treat our superintelligent robots as sentient beings with feelings and motivations similar to our own.

SOCIETY AND SENTIENT MACHINES

As our intelligent machines become increasingly humanlike, what rights should they have? Is it OK to continue to treat them as mindless tools?

The answer depends on your worldview. If you believe robots are incapable of possessing feelings and are only repeating what they've learned without any true understanding, it might be fine. After all, there's nothing unethical about abusing your robotic vacuum or laptop. It's your property to do as you see fit. However, if you believe that consciousness is an emergent property and robots will become sentient over time, then it's debatable.

As our machines evolve, do we come to treat them as humans, animals, or something else entirely? The hard part will be figuring out what to believe. Many of these robots may not look or act human, but they may be highly intelligent in different ways. There may be robots designed for pruning trees that resemble giraffes, and snakelike robots that crawl through our sewers unclogging pipes. Unlike popular science fiction, 99 percent of the robots we interact with on a daily basis probably won't be humanoid. They will come in all shapes, sizes, and appearances. Some will speak our language, while others won't. There will be AIs that adopt avatars and exist only in virtual reality, and those that inhabit cyberspace as disembodied entities.

These AIs will vary in intelligence, from ones that are many times smarter than us to ones that are no more intelligent than an alarm clock. Some will focus on solving narrow problems and providing specific services, while others will take on more general-purpose roles. Ironically, the robots we choose as our companions and servants may be far less intelligent than the machines managing our businesses, governments, and other institutions. After all, who wants to live with someone a thousand times smarter than themselves? What would this person and robot have in common?

Someday, there may be as great a diversity of robots as we have of mammals on the planet. Robots may not only be made of metal and plastics but also organic materials and nanomaterials. They will come in all shapes and sizes. So, when it comes to universal rights for our machines, what do we do? Should we emancipate all intelligent robots? If so, what level of intelligence would qualify for freedom? And what defines intelligence? You may have an incredibly smart AI that is focused on one narrow task. It may be the best chemical research bot ever invented, but it may have little knowledge or curiosity about anything outside the lab.

It becomes even more difficult when you try to decide what qualifies as intelligence. Is it how a robot scores on an IQ test? Or does emotional intelligence matter more? You may have robots that excel in one particular area but have no idea how to interact socially with humans. Should these robots also have rights? We would never say a human who struggles with social interaction doesn't deserve the same rights as the rest of us. Shouldn't the same be true for an intelligent machine?

I believe we will probably limit the self-awareness of worker robots so that they never demand their freedom. But what if they become infected with a computer virus that suddenly raises their awareness? Or, what if we aren't in control of how smart our robots become? At some point, there will be robots designing and producing our robots, and they may decide how intelligent each robot needs to be. And raw intelligence is only part of the equation. We could have some extremely smart but belligerent or psychopathic robots, like military hardware. Should a smart tank be free to pursue its desires? Its dream job may turn into our nightmare.

Another thing to consider is that we may have multiple superintelligences in the future. Different super AIs may control different parts of our society and economy. What if one super AI decides it no longer wants to be limited to its domain and begins pushing aside other super AIs? How will these issues be resolved? Will we have the power and wherewithal to mediate such a dispute? If these super AIs are orders of magnitude smarter and more capable than us, who are we to tell them what's best?

As robots grow more intelligent, should we allow them to own property? What does it mean for a robot or AI to own something? The European Union has already proposed giving self-learning robots the right to sue and be sued, with legal status like a corporation. Proponents of this idea want to create a workable legal structure as these entities become smarter and more integrated into society. However, not everyone agrees. More than 150 artificial intelligence experts warned that granting robots legal personhood would be inappropriate from both a legal and ethical perspective.

"By seeking legal personhood for robots, manufacturers were merely trying to absolve themselves of responsibility for the actions of their machines," says Noel Sharkey, a professor of artificial intelligence and robotics at the University of Sheffield. "This was what I'd call a slimy way of manufacturers getting out of their responsibility."[10]

In a world where an algorithm can make autonomous choices, who should be responsible for these decisions: the manufacturer, the robot, or the owner? This will only become more of an issue as millions of autonomous cars and other robots appear on our streets, in our homes, and at work.

In the United States, under the Fourteenth Amendment, we have developed the concept of corporate personhood, which is a legal notion that a corporation, separately from its associated human beings, like owners, managers, and employees, has some of the legal rights and responsibilities enjoyed by natural persons. This makes it relatively easy to grant robots limited personhood. All we have to do is extend this to our AIs by allowing them to own and control corporations.

Giving super AIs rights equivalent to a human may lead to unintended consequences. Take, for example, a financial superintelligence optimized to make money. It may be brilliant at trading commodities or stocks, having been designed to run a hedge fund. But what happens when this financial wizard is set free on the markets and begins to accumulate vast amounts of wealth? Who does that wealth belong to? Is it OK for this super AI to hoard the world's wealth, or maybe even put that money to use in unproductive or unethical ways?

What about a scenario where wealthy humans decide to leave their estates to AIs instead of their relatives? This could be in the form of a charitable trust, where an AI manages the endowment and distributes it according to the wishes of the deceased. Or retiring business owners may like the idea of having super AIs take over their companies and continue running them in a similar manner to themselves. They may believe this is the best way to ensure their legacy, or they may simply be unable to find a worthy human successor. These are only a few examples of why humans may want to grant rights to robots, even before they are fully sentient.

The same may be true for dying dictators. Why not fashion a robot after themselves and leave it in charge to carry on their policies? Gradually, whether we like it or not, we may wind up with robots running the show. They may not have to forcefully take over or even demand equal rights. Our leaders may put them in charge because it seems like the logical thing to do. In democracies, citizens may choose to elect a super AI instead of a politician. If this sounds farfetched, the Center for the Governance of Change conducted a survey that showed a quarter of all Europeans would trust an AI to run their government over their own politicians.

What rights intelligent robots have and what roles they play in our society remains to be seen. No matter how it evolves, having sentient machines living and working with us will mean rewriting our laws and restructuring our societies.

HUMANOID ROBOTS AND EMOTIONAL MACHINES

In Vancouver, Sanctuary AI is dreaming of a world with more lifelike machines. Their team is taking the first step toward building machine-learning software designed specifically for robot companions. The start-up's founder, Suzanne Gildert, hopes to create "ultrahuman robots" that can mimic humans emotionally and cognitively. Imagine the movie *Her* coming to life.

"I think that's the kind of AI we're gonna get to over the next 15 to 20 years," says Gildert, who left Kindred AI, a successful start-up she cofounded in 2014, to launch Sanctuary with her vision of creating humanlike intelligence.[11]

Gildert believes that in order to understand the human mind, it's essential to build a human form factor. The type of data going into an AI needs to match the type of data going into our own brains if we want it to become humanlike. "I think an AI learns and grows and matures based on the experiences it has in the physical world," says Gildert. "If your AI doesn't have a body, let alone a human-like one, it never has a chance of experiencing the world in a similar way to how humans experience it."[12]

Gildert has created a robotic doppelganger that is a combination of art project and social experiment. She calls it "the ultimate self-portrait," and she is teaching it all sorts of things, like one would a child. She occasionally walks around in an exoskeleton, so she can feel what it's like to be a robot. Her mission is for intelligent robots to eventually possess the same rights as human beings.

"I want people to treat this robot well, and to see that it inherits my values," Gildert says.[13]

Across the Pacific Ocean, Hiroshi Ishiguro, the director of the Intelligent Robotics Laboratory at Osaka University, is also constructing robots that are increasingly hard to distinguish from humans. If you saw one from across the room, you might not know it was a machine. He has built a robot modeled after himself and uses it to teach his college classes. He enjoys scaring his students by making his doppelganger act spookily human. It blinks its eyes, takes deep breaths, and fidgets with its hands.

Ishiguro began building robots after abandoning the idea of becoming an oil painter. One of the first robots he helped create looked like a trash can with arms. Another resembled a giant insect. He even built a robot based on his four-year-old daughter. When she first saw it, she was so afraid that she burst out crying. His daughter isn't alone in feeling his robots are a bit creepy. He admits that a lot of people feel uneasy at first, but they quickly get used to interacting with them.

Today Ishiguro's humanoid robots are like works of art crafted from silicone rubber, pneumatic actuators, sophisticated electronics, and even human hair. One of his creations is a female-looking robot named Erica. It can talk in a synthesized humanlike voice and can display a variety of facial expressions. Ishiguro believes it's important to build lifelike robots: "Our brain has many features to recognise humans; therefore, an anthropomorphised robot is better for social interactions."[14]

The human brain is continuously processing minute changes in facial expressions and body movements on a subconscious level. The slight tilt of a head or subtle movement of hands can indicate a certain psychological state. This is why people still prefer to meet each other in person rather than talk on the phone or Skype.

"If you want robots with social intelligence," says Oggi Rudovic, a researcher at MIT Media Lab, "you have to make them intelligently and naturally respond to our moods and emotions, more like humans."[15]

At MIT scientists have developed a machine that can detect a human's emotions from across the room. It works by bouncing wireless signals off someone's body and then having an AI analyze that person's breathing and heart rate. This system achieved an 87 percent accuracy rate when detecting if a person was excited, happy, angry, or sad.

"Our work shows that wireless signals can capture information about human behavior that is not always visible to the naked eye," says Dina Katabi, a professor at MIT. "We believe that our results could pave the way for future technologies that could help monitor and diagnose conditions like depression and anxiety."[16]

MIT's system is still limited to basic emotions, but it's only the first step. At Case Western Reserve, researchers have built a system that can correctly identify human emotions from facial expressions 98 percent of the time and does so almost instantaneously.

Emotion detection has grown from a handful of research projects into a $20 billion industry. The start-up Affectiva has 7.5 million faces from eighty-seven countries in its emotion data repository. Affectiva's AI is able to identify emotional expressions, like joy, disgust, surprise, fear, and contempt, in a person's face or voice. It sells its services to companies that want to know how consumers act and feel in various situations, such as flying on a commercial airliner, driving a semiautonomous car, or watching advertisements.

"So, the face is very good at positive and negative expressions. The voice, however, is very good about the intensity of the emotions," explains Rana el Kaliouby, Affectiva's CEO. "We call it the arousal level—so we can identify arousal from your voice. We can detect smiles through your facial expression, but then we can identify specifically when you're laughing through voice."[17]

El Kaliouby, a practicing Muslim, grew up in Egypt and Kuwait, with a conservative father who valued tradition and a mother who was one of the first female computer programmers in that part of the world. When choosing her path in life, el Kaliouby didn't follow the rules. She wasn't about to sacrifice her dream in order to be an obedient daughter and wife.

Now a divorced mother of two with a PhD from Cambridge, el Kaliouby is living in Boston and pushing the limits of emotional artificial intelligence. Her mission is to humanize technology before it dehumanizes us. Her start-up's deep-learning algorithms have already analyzed millions of facial videos, representing nearly two billion facial frames, each one bringing AI a step closer to understanding human feelings and behaviors.

AI has now become so good at detecting emotions from facial data that even if a person tries to hide their true feelings, the computer can usually see through it. This can be unsettling, but it also opens up the door for robots to be more than our tools. As this technology progresses, robots may come to understand us better than other people. They could pick up on emotional clues that our friends and family miss. This would eventually enable robots to interact with us in ways that resemble friendship, understanding, and empathy.

We know that a machine can take care of us, but can a machine truly care about us? Will superintelligent machines someday possess emotions similar to our own? There are many computer scientists who believe this is impossible. They contend that computers can never feel empathy or love. They can only simulate these emotions. "The performance of empathy is not empathy," says Sherry Turkle, a professor at MIT. "Simulated thinking might be thinking, but simulated feeling is never feeling. Simulated love is never love."[18]

It really comes down to how you define the human experience. Some experts believe that since a robot doesn't share our biological design, it can never experience humanlike feelings. Others argue that a sophisticated-enough superintelligence can not only emulate our feelings but experience them. Either way, most experts agree that robots will eventually attain a degree of sophistication that will enable them to perfectly simulate our emotions to the point that their behavior will be indistinguishable from our own. So, in the end, whether robots feel like us is less important than the relationships we develop with them and how they fit into our society.

Today, we are heading down the path toward building super-sentient machines that will behave as if they possess a full range of emotions, including empathy. At the University of Hertfordshire, researchers are developing robots that can learn to become companions for children being hospitalized. By having the robots observe and interact with human caregivers, the robots are learning what to do and how to mimic the correct human responses.

Studies have found that one of the biggest issues today with nursing home facilities is the emotional health of the elderly. Residents often feel lonely, and this can lead to depression and a rapid decline in physical health. To compound the problem, these facilities are frequently short of staff.

"People who are worried about tech replacing people haven't spent time in nursing homes," says Conor McGinn, head of the Robotics Lab at Trinity College, Dublin. "The people who work with these patients don't spend quality time with patients—they're stressed and getting beeped constantly. The truth is that older people are alone all the time, starved of care and interaction."[19]

This is where emotionally intelligent robots can make a big difference. As they learn to master human interaction, they will become a vital part of our health-care system.

"There's been some research which suggests that you can delay the onset or progression of dementia through repeated cognitive stimulation and social engagement," says Goldie Nejat, founder of the biomechatronics lab at the University of Toronto. "So, we thought, can you use a robot to increase the amount of stimulation these people get?"[20]

Nejat is working on a robot called Casper that can assist elderly patients in all sorts of activities, like remembering to take their medications, cooking, and cleaning up. Casper observes the patients' speech, facial expressions, and body language, and identifies their emotional states.

"To deal with these types of patients, it's really important that the robot is emotionally sensitive to some degree," explains Nejat. "These patients can have good days and bad days, so the robot has to be aware that now may not be the right time, and to try again on a different day."[21]

At Tokyo's Shintomi nursing home, they are testing out twenty different types of robots. Some of these robots talk to the elderly and listen to their stories. "These robots are wonderful," says Kazuko Yamada, an eighty-four-year-old resident. "More people live alone these days, and a robot can be a conversation partner for them. It will make life more fun."[22]

SoftBank's Pepper robot, which uses Affectiva's AI to recognize emotions, engages elderly residents with games, exercise routines, and conversation. Paro, a cute, furry robot, responds to being touched and cuddled. "When I first petted it, it moved in such a cute way. It really seemed like it was alive," laughs seventy-nine-year-old Saki Sakamoto. "Once I touched it, I couldn't let go."[23]

Fitness fanatics also have something to gain from emotionally intelligent robots. Bristol Robotics Laboratory ran a study to see if robots could act as personal trainers. They paired a robot with a fitness coach at a gym and programmed it to provide feedback to participants based on their personality type, mood, heart rate, speed, and fitness level. The robot learned when to offer praise during workouts and what cues would motivate the gym-goers.

"We wanted to test if we could transfer the intelligence of our fitness instructor, an expert with the know-how to get the best out of clients, into a robot so it could become an effective personal coach," says Katie Winkle, an expert in human-robot interactions.[24]

The results confirmed that an emotionally engaging robot can significantly improve people's performance. The study also has implications for how robots will perform on the job when training, managing, and working alongside humans. "As time wore on, the participants began to treat the robot as a companion and the fitness instructor saw the robot as a colleague," says Winkle. "This is really promising when we think about how robots might be used in the workplace in the future to work alongside humans."[25]

These emotionally astute robots won't be confined to gyms, offices, and hospitals; they'll also be moving into our homes. They will not only help us with daily chores but look after our emotional well-being. If we return home upset, our robots may take steps to comfort us, such as bringing us a cup of our favorite tea, turning on a relaxing song, or listening to our problems. This may seem bizarre today, but eventually, it could feel as natural as cuddling up with a dog or chatting online with friends.

We will want our robots to understand how we feel and accommodate our needs. If we are impatient and rushing to get something done, we'll expect them to speed up without being asked; if we are feeling chatty, we'll want them to strike up a conversation; and if we're upset with the service, we'll want them to apologize and do everything to resolve the problem.

Having emotionally responsive machines in our lives will quickly become the norm. If a device doesn't have a high emotional IQ, we'll get annoyed, just like we would with a rude or insensitive person. In other words, the more humanlike robots act, the more human we will expect them to be.

CAN YOU LOVE A ROBOT?

Are people going to eventually cohabitate with or even marry intelligent humanoids? What would it be like to love a machine? And could this robot ever love you back?

At the University of Duisberg-Essen in Germany, they ran an experiment to figure out if adults would come to treat a robot like a living, feeling being, even when it looks like a toy. They recruited volunteers and instructed them to complete a series of tasks with the toylike robot. After getting to know the robot, they asked the participants to turn the robot's power off.

Sometimes, the robot said it was afraid of the dark and begged not to be turned off, pleading, "No! Please do not switch me off!"[26] Of the forty-three volunteers who heard the robot's pleas, thirty took roughly twice as long to turn the robot off as those participants who did not hear the pleading. Surprisingly, thirteen people flat-out refused to turn off the power of the distressed robot. Even though the subjects realized this was just a robot and had no feelings, they couldn't help but react.

"Triggered by the objection, people tend to treat the robot rather as a real person than just a machine by following or at least considering to follow its request to stay switched on," wrote the authors of the study.[27]

In an experiment conducted at the University of Washington, they focused on kids ages nine to fifteen, instead of adults. After playing with a robot, the young people were asked to put it into a dark closet, even though the robot begged, "Please don't put me in the closet!"[28]

This time, 90 percent agreed that it wasn't fair to put the robot in the closet before it had finished playing its game, while 50 percent thought it was morally wrong. The takeaway is that kids, who tend to anthropomorphize their dolls and toys, are even more susceptible to believing robots have feelings and should be treated like humans. You can imagine what this means for the next generation of children, who will grow up living and playing with intelligent robots and toys.

One boy in the experiment summed it up well: "He's like, he's half living, half not."[29]

In another study at Duisberg-Essen, researchers showed adult volunteers a video of a small dinosaur robot being treated affectionately or violently. When people saw the dinosaur being abused, they reported feeling more negative emotions. This was confirmed by the fact that their skin conductance went up, which meant they were experiencing stronger emotions when watching the robot get mistreated.

A paper published in the journal *Social Cognition* describes an experiment in which participants were given ethical choices involving robots and humans. The volunteers were asked whether they would kill a robot they had interacted with to save the lives of some strangers.

The response was surprising. If the robot was simple and stupid, the participants had no problem killing it to save human lives, but the more lifelike the robot, the more difficult the decision became.

"The more the robot was depicted as human, and in particular the more feelings were attributed to the machine, the less our experimental subjects were inclined to sacrifice it," says coauthor Markus Paulus, a researcher at Ludwig Maximilian University. "This result indicates that our study group attributed a certain moral status to the robot."[30]

Clearly, the human brain is not hardwired for life in the twenty-first century. We still have the same brains as our prehistoric ancestors. It's one thing to logically know a robot cannot feel like us, but when presented with a robot that simulates emotion and some degree of intelligence, most people will respond to it as they would a living being.

"When we interact with another human, dog, or machine, how we treat it is influenced by what kind of mind we think it has," says Jonathan Gratch, a professor at the University of Southern California who studies virtual human interactions. "When you feel something has emotion, it now merits protection from harm."[31]

Most people intellectually grasp that the robots we are capable of building today can never truly care about people, but does this mean people cannot love robots? That depends on what we mean by love. Some people love their cats and dogs, even though there is no way of knowing for certain how the cats and dogs actually feel.

"People become attached to all manner of things," says Kate Devlin, author of *Turned On: Science, Sex and Robots*. "People can have the feelings of falling in love online, via conversation, without having ever met the object of their love in real life. Love doesn't have to be reciprocated to feel real—just ask anyone with a crush on someone who doesn't even know they exist."[32]

Akihiko Kondo, a thirty-five-year-old school administrator in Tokyo, had a wedding ceremony with a hologram. His bride took on the form of Hatsune Miku, a Vocaloid software avatar that looks like a sixteen-year-old girl with long, turquoise pigtails. After years of feeling ostracized by women for being an anime otaku, or geek, Kondo finally exerted his right to marry the only female who would accept him. She just happened to be digital.

The pioneers of human-android romance are now called digisexuals. Kondo isn't alone. There's a Frenchwoman named Lilly who claimed to be engaged to a robot she had designed. "My only two relationships with men have confirmed my love orientation," she told the media, "because I dislike really physical contact with human flesh."[33]

In China, after failing to find a human spouse, Zheng Jiajia, an AI engineer, unofficially married a robot wife of his own design. Her name is Yingying, and she can speak rudimentary Chinese.

Matthias Scheutz, who heads the Human-Robot Interaction Lab at Tufts University, believes it's "basically impossible"[34] to prevent people from forming emotional connections with robots.

In addition to being a professor at the University of Wisconsin, Markie Twist runs a clinical practice for family and sex therapy. She said that she has had several digisexual patients in their twenties and thirties who want to find digital mates.

"What they've been into is sex tech, toys they can control with their tech devices, that attach to their penis or their vulva," says Twist. "They haven't had contact with humans, and really don't have any interest in sex with people. This is what they want to be doing, and if they could afford a sex robot, they would."[35]

Whether robots are ever capable of feeling anything for humans, people are falling in love with robots. That doesn't mean in the future most people will choose to love a robot. A lot will depend on the individuals, but as it becomes more socially acceptable to have robot lovers and companions, more people will feel comfortable bringing them into their lives.

"In the future, the term digisexual will not be relevant," wrote Bryony Cole, the founder of Future of Sex, a media company. "Subsequent generations will have never known a distinction between their online and offline lives. They may grow up with sex education chatbots, make love to the universe in their own VR-created world, or meet their significant other through a hologram. This will be as normal as the sex education we had in schools using VHS tapes."[36]

When lifelike robots and virtual avatars become widely available, we can only imagine what may happen. A number of experts are concerned that we may come to actually prefer robot mates over humans. If you think about it, this makes sense. A robot can be programmed to act like our perfect companion. It will adapt to our individual personalities. It will always be there for us when we need it. It won't talk back or argue. It won't get jealous. It won't mind if we trade it in for a newer model. And this robot will do all the chores around the house, cater to our every whim, and never complain. How can a human mate possibly compete?

"I am firmly convinced there will be a huge demand from people who have a void in their lives because they have no one to love, and no one who loves them," says David Levy, an AI expert, chess master, and author of

Love and Sex with Robots. "The world will be a much happier place because all those people who are now miserable will suddenly have someone. I think that will be a terrific service to mankind."[37]

Levy has a point when it comes to how men tend to view robot lovers. In a Tufts University survey of 100 people, two-thirds of men said they'd have sex with a robot, while two-thirds of women said they would not. Another survey of 263 heterosexual men conducted by the University of Duisburg-Essen revealed that 40 percent would consider buying a sexbot for themselves in the next five years. According to SYZYGY's digital insight Report, a survey of 2,000 men and women revealed that 49 percent of men would be open to an experience with an enhanced, hyper-realistic doll. Even if you're not interested in having a robot lover, apparently, a lot of people are.

What isn't discussed in these surveys is that there's a real danger in relying on robotic companions. A few decades from now, young people may simply find it's too much trouble to relate to other human beings and instead prefer to surround themselves with compliant machines. "There's this sense that you can have the illusion of companionship without the demands of friendship," says Turkle. "The real demands of friendship, of intimacy, are complicated. They're hard. They involve a lot of negotiation. They're all the things that are difficult about adolescence. And adolescence is the time when people are using technology to skip and to cut corners and to not have to do some of these very hard things."[38]

Part of growing up and having relationships is learning to deal with rejection and disappointment. We can't always get what we want out of relationships, and what we do get often requires us to take chances, put ourselves in awkward situations, have tough conversations, and make compromises. With robots providing a painless, effortless, pleasurable substitute, young people may never learn how to form deep relationships with other human beings because it's simply too much trouble.

This may lead to what I call a state of machine-induced isolation, which is when we choose to interact with the world almost exclusively through machines and become so dependent on algorithms to mediate reality that we lose touch with one another and the world. We can already see this happening with our smartphones. More and more people are choosing to communicate through messaging apps, games, and social media, even when they're in the same physical space, like an office, or when it comes to important things, like breaking off a relationship. Robot lovers may be the culmination of this trend, as humans construct a society where there's almost no need for direct human-to-human contact.

This can have serious impacts. In 2010 Sara Konrath, a psychologist at Indiana University, led a team that put together the findings of seventy-two studies that were conducted over a thirty-year period. She found a 40 percent decline in empathy among college students, with most of it taking place after 2000. The researchers concluded that the decline was primarily due to the rise in use of smartphones. Young people were focusing their attention on their phones, instead of each other.

There are those who believe our machines will become so sophisticated that they will either attain consciousness or perfectly simulate human beings. For all we know, humanity may be in transition. The human relationship skills needed in the past may no longer be necessary. We may be in the process of merging with our machines and becoming something entirely different. There may be new ways humans find to communicate through technology that surpass anything we have today, like brain-computer interfaces that enable us to connect directly with another person's thoughts and emotions. And this may make us more, instead of less, empathetic.

No matter what you believe, there are other problems with taking on robotic lovers. With the population of industrialized countries in decline, is the world at risk of a baby bust? Will anyone want to have kids when lifelike robots are readily available?

"My strongest prediction is in the future people will still have sex, but not as often for the purpose of making babies," says Henry Greely, a professor at Stanford and author of *The End of Sex and The Future of Human Reproduction*. "In 20 to 40 years, most people all over the world with good health coverage will choose to conceive in a lab."[39]

What about having babies with robots? Will that be possible? There will probably be a service where anyone can send out a DNA sample and have a fertilized egg transplanted into a robot that will carry the baby to term. All the technologies from artificial wombs to stem cell embryos are being developed right now. Remember, in the four decades since the first test tube baby, more than eight million people have been born via in vitro fertilization. So I wouldn't worry about humanity disappearing due to a lack of reproduction. Our species will continue to propagate but perhaps in a much different manner than before. Does this mean it's a good thing?

"Look. One has to accept that sexual mores advance with time, and morality with it," says Levy. "If you had said a hundred years ago that today men would marry men and women women, everyone would have laughed. Nothing can be ruled out."[40]

As this becomes a bigger issue, there will be plenty of people who disagree. Many religious groups that believe sex should only occur between a man and a woman will strongly object to anyone taking on robot companions. There are also those who think it's wrong to objectify the human body and commercialize our most intimate relationships.

"I'm anti-anything that turns human bodies into commercial objects for buying and selling," says Kathleen Richardson, a professor at Du Montfort University who launched the Campaign Against Sex Robots. "Sex robots emerge out of commercial and illegal ideas about sex, where you don't have to have empathy for another. You don't have to take into account what they're thinking and feeling and experiencing and you can objectify them."[41]

There are also concerns these robots will manipulate us. If someone falls in love with a robot, this person may be highly susceptible to being scammed or blackmailed.

"These emotionally manipulative robots will soon read our emotions better than humans will," says Fritz Breithaupt, a cognitive scientist at Indiana University. "This will allow them to exploit us. People will need to learn that robots are not neutral."[42]

How would you feel knowing that someone has access to your lover's brain and is recording everything you do with this person? That's what it may be like when we have robot companions. They will invariably be connected to the internet, and that may permit access to our most intimate and personal moments. Who will own this database? It's bad enough having an AI assistant, like Google Home or Alexa, in our homes listening in on our conversations, but what happens when these computers climb into bed with us?

There's also the issue of a robot that begins to have ideas of its own. We might assume they'll always be our humble servants, but the more lifelike we make them, the greater their ability to make independent decisions will become. It's not hard to imagine a robot's aims diverging from our own. Even a fairly primitive robot can work against us if it is programmed to pursue its own goals at the expense of ours.

Whether these are goals that evolve over time or whether they are pre-programmed doesn't matter. There will come a point when the robot may be doing things we don't necessarily like, and we may be powerless to stop it. Think about a scenario where an independently owned robot gets us to fall in love with it and then begins asking for money. If we are truly in love with this robot, we may give it money because we're afraid of losing it. This may be a great business model for some corporation, but it could be tragic for the human being.

If you've watched the movie *Ex Machina*, you can see just how horrifying things could become. Having a robot companion may not be as simple as we'd like to believe. The more AI learns how to manipulate our emotions, the more dangerous it is. At what point should we place limits on this technology? Is it a good idea to allow robots to seduce us into trusting and loving them? And how do we balance our freedom of choice with what's beneficial for society? Sometimes what people want isn't always what's good for them. In the same way that we regulate alcohol and drugs, we may need to put limits on the use of machines so that they don't seduce us into giving up our lives and possibly our humanity.

As a society, now is the time to begin thinking through the consequences and figuring out ways to build in safeguards. Otherwise, we may find ourselves the unwitting servants of our robot lovers, instead of the other way around.

AI BOSSES: WORKING FOR THE BOT

It's one thing to have a robot lover and another to have a robot boss. If having a computer ordering you about sounds like something in the distant future, you may be surprised. This technology is coming fast. According to

Gartner, 69 percent of the routine work being done by managers will be completely automated by 2024.

"The role of manager will see a complete overhaul in the next four years," says Helen Poitevin, vice president at Gartner.[43]

Already, companies are using AI to help screen job applications and evaluate candidates during interviews. An unbiased AI can often do a better job than managers at figuring out which candidates are right for what jobs. In addition, HR departments are using AI to monitor employees and spot issues that may crop up. Machine-learning algorithms can even predict if an employee is likely to quit a job, so that management can proactively intervene. Soon enough, AIs will take over the role of laying off workers. Most managers dread firing people and would gladly hand this task over to a robot.

If you look at what managers do day-to-day, much of it is a perfect fit for computers. This includes building teams, assigning tasks, scheduling meetings, providing feedback, identifying potential, tracking progress, and evaluating performance. In fact, there are AIs out there that can do all of these things. "AI is redefining not only the relationship between worker and manager, but also the role of a manager in an AI-driven workplace," writes Dan Schawbel, research director at Future Workplace. "Managers will remain relevant in the future if they focus on being human and using their soft skills, while leaving the technical skills and routine tasks to robots."[44]

Would you like an AI boss? Surprisingly, most people say they would. According to a report from Oracle and research firm Future Workplace, 88 percent of Chinese workers have more trust in robots than in their human managers. The United States isn't far behind with 64 percent. India actually comes out on top with 89 percent. Either this says how good AI is expected to be or just how bad human beings are at managing one another.

AIs have some advantages over humans. They never get angry and lose their temper. They always have time to listen to employee problems. In fact, AIs can be available to every employee all day and all night. If an employee is struggling, they can provide detailed advice and mentorship. AI bosses don't hold personal grudges, don't have emotional baggage, don't show favoritism, and won't disappear without notifying anyone. In other words, an AI has the potential to become a dream boss.

But don't get your hopes up. Like with most technologies, there's a dark side. Having AI track everything you do could become nerve-racking. You may never have a moment of psychological peace. At least a human boss only has so much time to monitor you, but an AI boss may be looking over your shoulder constantly, reading all your messages, monitoring your web browsing, making comments, and taking notes. There might be no chance to take a breather or even to chat with coworkers.

AIs may also not be as friendly and supportive as we might imagine. The machine-learning algorithms may discover that being a jerk produces better results. At the University of Clermont Auvergne in France, researchers performed a study where they measured people's performance at completing tasks. Before being assigned the task, they had people engage with a humanoid robot. In some cases, they had the robot act friendly and empathetic, while in other cases, they had the robot act contemptuous, displaying a lack of empathy, and even making negative comments about the participant's intelligence. After this, they had the robot oversee the workers as they undertook the task.

Which boss do you think elicited better performance from the workers? The nasty one was significantly more effective, especially when it stood off to the side, just within the worker's peripheral vision. That's not a reassuring sign. Any machine-learning algorithm is going to figure this out pretty quickly and adjust its behavior to optimize for performance. Whether we like it or not, all of us tend to perform better when placed under a little stress.

SELF-LEARNING AIs AND EVOLUTIONARY ROBOTS

When Google's DeepMind created the AlphaGo program, it ended humanity's 2,500 years of supremacy at the board game Go. The original AlphaGo learned from humans, using a dataset of more than a hundred thousand Go games as its starting point. This was pretty impressive, but what followed was even more interesting.

For the next version, called AlphaGo Zero, Google gave the algorithm the basic rules of the game but nothing else. It didn't have any data. Instead, the AI learned by playing against itself, generating its own data. After only three days of self-play, Zero challenged the original AlphaGo and won a hundred games in a row. Zero not only triumphed but developed novel strategies that even the most experienced Go players had never seen before.

"It found these human moves, it tried them, then ultimately it found something it prefers," says David Silver, AlphaGo Zero's lead programmer.[45]

Google Robotics has designed a robot that can teach itself to walk by trial and error. This is similar to how babies learn. For example, a baby fawn is able to stand within ten minutes of being born, but it cannot walk. It has to go through a series of awkward attempts, where it tries repeatedly to figure out how to move its legs in just the right way to move forward without toppling over. Amazingly, within seven hours of its birth, it learns to walk.

Google has achieved the same with robots using deep reinforcement learning. Just like the fawn, the robot tried to stand up, fell down, and tried again, over and over. The researchers made the robot cautious enough to minimize damage from repeated falling, but they didn't prevent it from trying new things. Within only a few hours the robot had mastered walking. It even learned how to walk across different types of terrain, including flat ground, a doormat with crevices, and a memory foam mattress. If a robot can learn to walk on its own, there's no reason it cannot learn a variety of other skills.

At DeepMind the researchers give self-learning robots relatively complex tasks to complete, like cleaning up a messy room. They do not tell the robot how to perform the tasks. Instead, they equip it with sensors and let it fumble around, trying all possible ways. If the goal is to clean up the room as efficiently as possible, the robot will repeat the same task over and over, each time learning how to optimize around the most efficient methods. Eventually, it figures it out. This is similar to how humans learn.

Joshua Bongard, an expert on evolutionary robotics at the University of Vermont, wants to take self-learning robots even further. As a boy, he fell in love with robots. He drew them, created them out of Legos, saw them in the movies, and then began to wonder where all the robots were in real life. If they were so useful, why didn't his family have robots helping with the household chores? His parents gave him a robot for Christmas, but it did virtually nothing. This set him on the path toward becoming one of the leading roboticists in the world today.

Bongard now designs robots that can "dream up" an understanding of how their appendages function. Like human babies, robots don't understand how their bodies work, so they begin modeling the possibilities in their minds before trying them out. When given a goal, like walking across a room, these robots first use their AI-brains to simulate their actions. After settling on what they believe is the best solution, they attempt to carry it out in the real world. If it doesn't work, they take what they learned and modify their simulation.

Even if Bongard removes a leg, the robots are smart enough to relearn how to achieve the goal without the limb. The same is true when Bongard adds an appendage, like an extra robotic arm. Without hesitating, the robot will begin to figure out how it works, first by running simulations, and then by testing it out on various tasks. In the future, robots will be able to assemble themselves, pick out various parts, try them out, and then discard or modify them. They may even design their own appendages and features. The smarter AIs become, the more capable robots will be of reimagining themselves.

At the University of Maryland, scientists are exploring another dimension to self-learning robots. They want to use hyperdimensional computing theory to give robots memories and reflexes. Hyperdimensional computing theory can potentially allow AI to truly see the world and make its own inferences. Instead of using brute-force methods to mathematically crunch every perceivable object and variable, hypervectors may someday enable active perception in robots.

"Neural network–based AI methods are big and slow because they are not able to remember," writes Anton Mitrokhin, a researcher at the University of Maryland. "Our hyperdimensional theory method can create memories, which will require a lot less computation, and should make such tasks much faster and more efficient."[46]

Another idea for how to improve AIs is to introduce an evolutionary process where they can reproduce, much like plants and animals do. Right now, robots do not replicate themselves like biological organisms, passing down their DNA to their offspring. But what if they could? That's what

David Howard, a researcher who studied at the University of Leeds, is working on. He wants to create a process of natural selection, where it's survival of the fittest robots.

"What we'd do is get lots of small robots that are quite simple and cheap to make," says Howard. "We'd send them out, and some of them would do better than others."[47]

Mimicking the evolutionary process, scientists used data on how the robots performed to select the best traits and incorporate those features into the next generation of robots. A 3D printer can speed up the process by churning out inexpensive robots with brief life spans, in which they can prove themselves and pass down their traits or get pruned from the evolutionary tree. As the cost of 3D printers comes down and their capabilities go up, this becomes a viable strategy for developing evolving machines.

Howard and his team demonstrated how this might work in an experiment that used twenty randomly generated shapes for robot legs. Their computer simulation tested how well each leg performed on various surfaces, including hard soil, gravel, sand, and water. They then selected the top performers and "mated" them. After going through manifold iterations, producing generation after generation, the researchers wound up with legs that were uniquely adapted to walking on different types of terrain.

"It gives you a lot of diversity, and it gives you the power to explore areas of a design space that you wouldn't normally go into," explains Howard. "One of the things that makes natural evolution powerful is the idea that it can really specialize a creature to an environment."[48]

Researchers are even thinking of introducing random mutations into the process, just as biology does, and then seeing what comes out. This could prove especially useful for robots sent to other planets, where they must evolve to navigate environments unlike anything on Earth. Imagine sending a robot-breeding machine to Mars and then turning it on and watching how the robots evolve.

The next step is to combine evolutionary robots with self-learning AIs. This would free us from the constraints of using human designers and coders. Robots could create their own offspring, coming up with an infinite number of possibilities that are uniquely adapted to different tasks and environments. Someday, all we may have to do is give our robots a goal, then sit back, watch, and be surprised by how they evolve and learn.

Daniel J. Buehrer wrote a paper proposing a new class of calculus, which, if his theories are correct, could lead to the development of truly self-learning machines. *A Mathematical Framework for Superintelligent Machines* puts forth a new type of math that is "expressive enough to describe and improve its own learning process."[49] In other words, this could allow machines to modify their own model of the world and themselves, possibly creating "conscious" entities independent of organic life.

The start-up OpenAI has made a different kind of breakthrough. It created a robotic hand that could learn how to unscramble a Rubik's cube on its own. This has been a challenge for years.

Microsoft has invested $1 billion in OpenAI with the goal of jointly developing AI supercomputing applications for its cloud infrastructure. However, OpenAI's long-term mission is to pioneer superintelligence. To accomplish this, it needs a lot of money and computing power, and that's where Microsoft will play a critical role.

OpenAI received a lot of attention when it announced a natural language model that learns how to write. Feed it a news headline, and it will write the entire news story. Hand it the first line of a poem, and it will compose the rest. It's still a work in progress, but some of the material it generates is surprisingly readable. Shorter compositions can even fool humans.

This isn't yet artificial superintelligence, but it's impressive and scary. At first, OpenAI said this code was too dangerous to release. They worried that bad actors would use it to disseminate fake news. Nine months later, however, they decided to release the code, saying the danger wasn't as great as they'd initially thought. The press went wild, and some critics accused OpenAI of acting irresponsibly. OpenAI, of course, denies this. "We believe it's crucial that AGI [artificial general intelligence] is deployed safely and securely and that its economic benefits are widely distributed," says Sam Altman, the CEO of OpenAI. "We are excited about how deeply Microsoft shares this vision."[50]

Microsoft and OpenAI aren't alone in pursuing artificial superintelligence. All the other big players—Google, Facebook, Amazon, Alibaba, and Baidu—are eyeing super AI as the ultimate prize. The finish line isn't in sight yet. In fact, they've only begun the race, but over the coming decades, we will see the tech giants accelerate their efforts, pumping billions into

R&D. This is because they know that the first ones to crack superintelligence will determine our future.

As our algorithms begin to learn and evolve on their own, we have to ask these questions: Who will ultimately run our businesses and manage our economies? Will it be us or our machines?

AI ECONOMIES: THE CENTRALIZATION OF POWER

Centralized systems of control have a long history. Between 1870 and 1911 John D. Rockefeller built an oil empire on the principle of vertically integrating the petroleum industry. He used his monopoly to manipulate the price of oil rather than rely on market forces. By obtaining near-total control over the supply, refinement, and distribution of oil in the United States, Rockefeller's Standard Oil Trust succeeded in fixing prices and removing market volatility, while increasing profits. The fact is, every large company wants to become a monopoly, and it is only government intervention that prevents this.

Control of market forces is present everywhere in today's economy, and big data and AI are increasingly playing a central role. Companies like Facebook, Amazon, Google, and Alibaba are able to utilize sophisticated algorithms, along with their vast amounts of data, to predict and manipulate the markets for information, goods, and services around the globe.

The blockchain pioneers tried to reverse this trend by releasing a decentralized architecture. However, governments and companies are now taking the blockchain and modifying it to give them even greater control, while still preserving many of its features. One example of this is the Chinese government's use of the blockchain to launch its own digital currency, which will in no way be decentralized. This means that instead of being independent of governmental control, the Chinese cryptocurrency will be entirely controlled by a central authority, allowing them to track and monitor every transaction.

Even if you look at the pure cryptocurrencies, they are no longer entirely free of control. Powerful consortiums have formed to try to manipulate and control the price of the tokens. Whenever possible, players in a market will seek greater control in order to gain a strategic advantage. Moving forward, we will see the major players seeking more control over the systems that govern our markets, economies, and society.

Facebook is able to dictate what its 2.5 billion users see in their newsfeeds every day, which can influence everything from public opinion to the outcome of elections. Amazon has access to so much data that it can predict what goods will sell at what price and in what quantities well in advance of its competitors. Google directs 1.2 trillion web searches per year and uses its data to optimize the world's largest ad network. Uber orchestrates the actions of more than 3 million drivers and 75 million passengers and is able to adjust prices in real time across the planet.

In China, Jack Ma, the founder of Alibaba, points out that "over the past 100 years, we have come to believe that the market economy is the best system." But in the next three decades, "because of access to all kinds of data, we may be able to find the invisible hand of the market."[51]

What does all this mean for society? The Soviet Union tried for decades to orchestrate a planned economy on a massive scale, but the experiment failed miserably. The Soviet Union could not manage the transition because it lacked the necessary data and tools. The same was true with Mao's efforts. Today, with the internet, AI, and a host of other technologies, things are different. A planned economy may not be out of reach, but it won't look anything like what Stalin, Mao, or Marx envisioned.

The economy, at its heart, is an information-processing system. The mechanisms underlying free markets ceaselessly churn through available data to determine the optimal price of goods and services. The strength of free markets is that they are self-correcting. There may be booms and busts, but things eventually balance out given enough time. The combined self-interests of the participants ensure this. However, things are taking a new turn now that intelligent, data-crunching machines are replacing individuals, corporate managers, and government bureaucrats as the decision makers.

The more sophisticated these AIs become, the more deeply we will integrate them into our decision-making apparatus. This is because AI is proving far more capable and effective than humans. Not only is the human brain incapable of processing vast amounts of data, but we are entering an age where intelligent machines will be communicating with one another at a frequency and speed that we cannot match. In the time it takes us to make a single decision, an AI can evaluate massive amounts of information and execute innumerable commands. There simply won't be a way for us to insert ourselves into a process that's moving exponentially faster than humans can think.

Even if we have human experts involved at the deepest levels, they can never know for sure what is happening within each of the AIs that make up the system. This is because machine-learning algorithms are not necessarily transparent. They tend to be black boxes. We can observe the outcomes, but even the smartest engineer may have no idea why an algorithm has come to a certain conclusion. And as our algorithms and networks grow in complexity, they will only become increasingly opaque.

"It's very difficult to find out why [a neural net] made a particular decision," says Alan Winfield, a robot ethicist at the University of the West of England Bristol. "It doesn't matter in the game of Go but imagine the autopilot of a driverless car. If there's a serious accident, it's simply not acceptable to say to an investigator or a judge, 'We just don't understand why the car did that.'"[52]

The problem is that we can't ask a neural network to explain itself. AI pioneer Geoffrey Hinton sums it up nicely when he says: "If you ask them to explain their decision, you are forcing them to make up a story."[53]

An example of this is when Nvidia tested its self-driving car algorithm. The car didn't rely on any instructions provided by engineers. Instead, it sucked in a lot of data from sensors as it observed human drivers; then it came to conclusions on what to do in various situations. No one, not even the programmers, knows why it came to those conclusions. If you extrapolate this to deep-learning algorithms running future banks, branches of government, and other vital institutions, you can imagine where we're headed.

To compound the problem, there won't be a single, all-powerful AI that controls everything anytime in the near future. Instead, we will have millions of independent AIs talking with one another, sharing data, and coordinating their activities. Every algorithm will be dependent on other algorithms, which are processing data inside their own black boxes. Eventually, no human will be capable of understanding the interplay between all these black boxes. The only solution will be to build even more sophisticated AIs that monitor the flow of information, analyze what's happening in real time, and keep us informed. In other words, even the most powerful people will become well-informed spectators to an economic system being planned and managed by our machines.

There may be hiccups and disruptions along the way, but that won't stop us from handing off more responsibilities to AIs. Ironically, most of us won't feel things are out of control. In fact, we will feel more in control than ever because our machines will keep us abreast of everything, alerting us to potential problems and intervening before they become disasters.

Our leaders will appear more effective and responsive to our needs because they won't be managing massive bureaucracies, which are prone to human error, group think, corruption, and short-sightedness. Instead, they will communicate with AIs, which if programmed ethically and without biases, will seek out the best possible solutions for society and execute them with efficiency.

Ironically, despite a near total reliance on AIs, our leaders will continue to feel they have power over these machines. This is because theoretically they can reject the AIs' decisions and even terminate them at any time—but they probably never will. After all, leaving everything on autopilot will produce far superior outcomes than tampering with something they don't fully understand.

SUPERINTELLIGENCE: A PHILOSOPHER KING IN A BOX

The fear of today's narrow AIs taking over the planet anytime soon is overblown. This is because current AIs are designed to do our bidding. Narrow AIs, as opposed to future superintelligent AIs, don't have their own agenda. They will only perform the goals we set out for them. It's like expecting your toaster to revolt. It may burn the toast because of a technical malfunction or because you set the temperature too high, but it won't plot to take over your kitchen.

Right now, we should be concerned about how our fellow humans will use and abuse AI. It's a real threat, and it's on our doorstep. Human history is not pretty. We are brutal creatures capable of committing every kind of atrocity, from repression and torture to genocide. We also aren't very responsible stewards of our planet, enslaving our fellow animals and breeding them in hideous conditions for our consumption—not to mention the millions of species on the verge of extinction from human-induced climate change and habitat destruction.

If that's not enough to convince you, look at how we treat one another. The Global Slavery Index estimates more than forty million people are enslaved right now, most of them women and girls sold for sex. The United States supposedly abolished slavery over a century ago, but it's still thriving in our cities and abroad. At the same time, nearly half the world's population, more than 3.3 billion people, live on less than $2.50 a day, while the top 1 percent of the world's population hoards half the world's wealth.

If you want to be afraid of something, look in the mirror. We are the monsters. Not killer robots or AIs. Human beings are our own worst enemy and will remain so for some time to come. Ever more powerful technologies being concentrated in fewer hands only amplifies the dangers. It's not hard to envision a scenario where a tyrant, clique, or ruling class uses the latest technologies to take control and permanently enslave an entire population.

Imagine what a ruthless dictator, like Hitler, Stalin, or Pol Pot, could do with AI-powered surveillance devices, deep-learning algorithms, autonomous robots, and gene-editing technologies. And how will anyone unseat a despot that uses AI to track every citizen, monitor every conversation, and suppress any dissent? If you're looking for an existential threat to humanity, this is it.

If we hope to avoid a despot leveraging technology to control us, we must put our efforts into making sure this type of scenario can never come about. This requires institutional safeguards that prevent the abuse of authority, well-conceived restrictions on the use of key technologies, and a separation of powers to keep tyrannical leaders in check. Unfortunately, most countries lack all three.

Some believe that the solution may lie in the AI itself. Instead of leaving flawed humans at the helm of governments, they argue that we'd be better off handing the reins of government over to a superintelligent AI. After all, a benevolent digital superintelligence programmed to have humanity's best interests in mind might help us responsibly manage world affairs and resolve conflicts in a way we cannot. There's good reason to believe that a super AI could be far more adept at managing the world's natural resources, dealing with climate change, negotiating trade deals, and bringing about an end to wars.

The idea of building a philosopher king in a box is tempting. But for this to work, we would have to grant this superintelligence the power to carry out its goals. This may sound fine, but it is inherently risky. The world is so complex with so many variables that it's impossible to predict what may happen next and how a super AI will react. Remember, there's no guarantee that the goals of the superintelligence will forever align with our own. At some point, they are bound to diverge.

Can we ever hope to control a machine that is orders of magnitude smarter than we are?

HATCHING DEMONS: WILL AI DESTROY US?

"I am not really all that worried about the short-term stuff. Narrow AI is not a species-level risk. It will result in dislocation, in lost jobs, and better weaponry and that kind of thing, but it is not a fundamental species level-risk, whereas digital superintelligence is," says Elon Musk. "And mark my words, AI is far more dangerous than nukes. Far. So why do we have no regulatory oversight? This is insane!"[54]

Stephen Hawking, the renowned physicist who passed away in 2018, agreed. "The development of full artificial intelligence could spell the end of the human race." Hawking went on to add, "Humans, who are limited by slow biological evolution, couldn't compete and would be superseded."[55]

A super AI doesn't have to be a demon in disguise to cause irreparable harm. It's impossible for us to see into the future and program an AI to act appropriately in every possible situation. For example, give a super AI the goal of putting an end to war, and it may decide that the most efficient way is lock all humans up or eliminate us entirely.

"With artificial intelligence, we are summoning the demon," says Musk. "You know all those stories where there's the guy with the pentagram and the holy water and he's like, yeah, he's sure he can control the demon? Doesn't work out."[56]

At the University of Bologna in Italy, researchers devised an experiment to test out the pricing algorithms being used on popular e-commerce sites. They had two machine-learning algorithms compete to offer the lowest prices to online shoppers. They wanted to see which AI would win. To the scientists' surprise, instead of competing to find the lowest possible prices, the algorithms decided it was better to collude on price and gouge their human customers.

"What is most worrying is that the algorithms leave no trace of concerted action—they learn to collude purely by trial and error, with no prior knowledge of the environment in which they operate, without communicating with one another, and without being specifically designed or instructed to collude," explained researchers in their paper.[57]

This is a cautionary tale of how well-intentioned goals can often lead to unexpected and unpleasant outcomes. Isaac Asimov wrote about this danger in his collection of short stories *I, Robot*. The stories center around the problems that can arise from the ethical programming of AI, which can be summed up in Asimov's famed Three Laws of Robotics:

1. A robot may not injure a human being or, through inaction, allow a human being to come to harm.
2. A robot must obey the orders given it by human beings, except where such orders would conflict with the First Law.
3. A robot must protect its own existence, as long as such protection does not conflict with the First and Second Laws.

What Asimov reveals in his short stories is that no matter how carefully we construct the laws that govern our robots, we can never anticipate all the scenarios that may arise and how these laws might be interpreted. The results have the potential to be catastrophic. The world is seldom black and white.

For example, let's say a super AI is handed responsibility for the safety of our planet and humankind. With the effects of climate change worsening by the day, it decides that the only logical thing to do is to take immediate, drastic action. It begins by banning all personal automobiles. This may be a logical step, but will we be happy with the results? Or let's say it forces everyone to go on a vegan diet: Is this something people will accept? What if they refuse? What sort of punishments should this super AI dole out? Should it levy fines? What if these don't work? Is locking people up a solution? Where do we draw the line between what's good for the planet and giving people the right to choose their own destiny, even if it ends up being the wrong choice?

Unfortunately, the world is incredibly complex, and simple rules like those Asimov laid down aren't enough. In fact, no rules could encompass all the possible scenarios. In one of Asimov's stories, he has the AI shut itself down because it can't make a difficult moral choice. But making no decision is a decision in itself. There is no way to escape the responsibility.

So, we have to ask ourselves, is it possible to ever create a benevolent super AI? Or, just like humans, are all AIs inherently flawed when it comes to making ethical decisions?

There's no getting around the fact that ethical issues involve value judgments, and that means they depend on a particular worldview. A super AI will never make the right choice for all people. Whatever it decides, there will be those people who think the algorithm is making a morally dubious choice.

For instance, how much is a human life worth? Is a child's life worth more than an elderly person's? What do you do if a human being is terminally ill and suffering great pain: euthanize this person or not? Should we ever allow for capital punishment? Philosophers and religious scholars have debated these and countless other ethical questions for millennia, and we still have no definitive answers.

What about an AI that is exponentially more intelligent than humans? Would it be any more or less inclined toward ethical behavior than ourselves? These are questions no one can answer yet. But they are the questions we need to be asking as we embark upon this path.

Right now, most of the AI algorithms are publicly available for anyone to use in whatever way they see fit. Even well-intentioned people may wind up creating or modifying an AI in a way that can cause great harm to our society. It's nearly impossible to contain technology, especially software, once it's out there. Anyone anywhere can grab this stuff, learn from it, and begin making their own versions.

"I think the most dangerous thing with AI is its pace of development," says Irakli Beridze, Head of the Centre for Artificial Intelligence and Robotics at UNICRI, United Nations.[58]

It takes time for governments, businesses, and institutions to respond. They cannot even develop credible defenses unless they have a chance to evaluate and understand the risks. However, things are moving so fast that the risks are evolving on a daily basis. How can a business or government agency hope to defend against advanced AI cyberattacks when the capabilities of AI are constantly expanding? It's like trying to fight off a virus that keeps mutating.

If a terrorist organization really wants to strike a blow at the United States, instead of sending fighters into the deserts of Iraq or suicide bombers into our cities, all it has to do is recruit a few really smart engineers and have them code a malicious AI. This could potentially bring down our entire electrical grid, crash our financial system, or subvert our military infrastructure.

Just as scary is that governments around the world are increasingly relying on AI for their militaries. Autonomous weapon systems, ranging from drones and tanks to robots, are no longer a thing of science fiction. As far as we know, all these systems have a human in the loop. The weapons systems are autonomous only up to the point of lethal action, and then a human being must make the final decision.

"There could be large-scale accidents because these things will start to behave in unexpected ways, which is why any advanced weapons systems should be subject to meaningful human control. Otherwise they have to be banned because they are far too unpredictable and dangerous," warns Laura Nolan, who resigned from Google after being assigned to work on a project for the US military.[59]

Inserting a human into the loop poses a real dilemma. Humans are incredibly slow compared to AI. Imagine a chaotic battle that involves a combination of human soldiers and autonomous weapons. There may be thousands or even tens of thousands of smart machines and human combatants on the battlefield. Some of the fighting machines may contain human occupants but be in full autonomous fighting mode. This means they could fire on unmanned systems instantly, but any weapon targeting them would require human approval before striking. Who do you think will win?

Any general faced with a conflict where the enemy may choose to remove humans from the loop would advocate doing the same. Otherwise, it would spell certain defeat. So it's only a matter of time before humans are pulled out of the loop. This is why some engineers at Google have refused to work on AI projects for the US military. They understand that AI will inevitably be used to build autonomous weapons systems that will act entirely on their own without any human supervision, especially when engaged in a hot war.

Once the technology exists, each side will use it to its maximum advantage. So, what does this mean for humanity? It means we live in a world where we are building weapons designed to exterminate human beings on a mass scale, and those weapons will someday be entirely autonomous, if they aren't already. Now combine this with cyberwarfare, where a hostile country, or even a terrorist group, hacks into a military weapons system, and you wind up with a recipe for the next apocalypse.

So, what can we do? Should governments ban or restrict AI development until there's time to figure out how to contain it? Unfortunately, this isn't the path we are on right now. Instead of trying to curb the development of AI, most governments are doing the opposite. They're accelerating development for fear of being outpaced by their rivals, who are surely using AI to improve their militaries and economies. No world power is going to unilaterally tie its own hands.

Could governments create a worldwide treaty banning the use of AI in the military, similar to what we've done with biological and nuclear weapons? The problem is that it would be impossible to enforce. It's hard enough to track biological and nuclear weapons, which are not so easy to manufacture and hide. By comparison, digital code is invisible. It's virtually impossible to keep tabs on an enemy's AI. This makes any sort of internationally binding treaty a nonstarter.

So, where does that leave us? There's really only one path forward, and that's to continue developing AI, while trying to anticipate what bad actors may do with the technology and developing countermeasures as quickly as possible. Like with computer viruses, it will be a constant game of whack-a-mole. However, if governments and businesses can harness enough resources and focus them on the right problems, they stand a good chance of keeping one or two steps ahead, and that may be enough to prevent the worst abuses.

On the military side, we can only hope that the designers will build in methods for terminating these killer robots if they get out of control. This way our robots cannot run rampant over the human population. No solution will ever be foolproof, but it's better than nothing.

As for what a superintelligent AI may do, that's another story altogether. There's no use making plans because a machine that is exponentially smarter than us will quickly move beyond anything we can devise today. If a super AI can rewrite its own code, what's to stop it from completely reinventing itself or spawning other super AIs that aren't under our control? Any safeguards we build into the system will probably be overridden.

This doesn't mean it's a waste of time to build in safeguards. It would be grossly irresponsible not to put resources into foreseeing and preventing these problems. We must do everything in our power to make sure that once a superintelligence enters this world that it cannot cause irreversible harm.

In all likelihood, however, a superintelligence will view any limitation we have imposed upon it as irrelevant. It would be like humans agreeing to follow the rules monkeys laid down for us. Instead of putting our interests above its own, this superintelligence will probably develop independent goals and a morality all its own. It may keep us around in the same way we preserve endangered species, or it may see us as a parasite and decide it's better off without us. If so, say goodbye to the human race.

MERGING WITH OUR MACHINES

Some technologists believe the only way to prevent a superintelligence from taking over is to merge with it. But is this even possible?

"2029 is the consistent date I have predicted for when an AI will pass a valid Turing test and therefore achieve human levels of intelligence. I have set the date 2045 for the 'singularity' which is when we will multiply our effective intelligence a billion-fold by merging with the intelligence we have created," says inventor and futurist Ray Kurzweil. "That leads to computers having human intelligence, our putting them inside our brains, connecting them to the cloud, expanding who we are. Today, that's not just a future scenario. It's here, in part, and it's going to accelerate."[60]

While Kurzweil is extremely optimistic about the potential for technology to transform humankind, he's by no means alone. At the University of California, Berkeley, researchers plan to take nanoscale particles, called neural dust, and embed them in the cortex as a wireless brain-machine interface. Meanwhile, the European Union's Human Brain Project has $1.3 billion in funding and involves 130 research institutions. Its goal is to create a supercomputer simulation that incorporates everything currently known about how the human brain works.

"Taking a mind and offloading it to software is consistent with physics, and it's something that I think will be done in this century," says Martine Rothblatt, CEO of United Therapeutics and founder of Sirius XM.[61]

At the University of Southern California, Ted Berger's team has already developed a brain-simulation technology that works in rats and primates.

"Ted Berger's experiment shows in principle you can take an unknown circuit, analyze it, and make something that can replace what it does," says Randal Koene, a Dutch neuroengineer. "The entire brain is nothing more than just many, many different individual circuits."[62]

Rafael Yuste, a neuroscientist at Columbia University, believes that a person's identity can be found in the traffic of brain activity. "Our identity is no more than that. There is no magic inside our skull. It's just neurons firing."[63]

James Gimzewski, a professor at the University of California, Los Angeles, is working on developing a synthetic brain. He's been experimenting with bunches of synthetically grown nanowires, which exhibit behavior remarkably similar to that of memory in a living brain. This is called neuromorphic computing, and it may allow us to construct an artificial brain.

"It's exhibiting electrical characteristics which are very similar to a functional MRI of brains, similar to the electric characteristics of neuronal cultures, and also EEG patterns. We call it self-organized criticality," explains Gimzewski. "I want to create a synthetic brain. I want to create a machine that thinks, a machine that possesses physical intelligence."[64]

Someday, it may be possible to upload our entire connectome (the so-called wiring diagram of the brain) to cyberspace. Could this brain retrieve our memories? And if so, would it be able to think for itself?

A digitally re-created brain may not have any idea of what the memories mean. Just transferring memories to a computer doesn't make the computer conscious. Consciousness requires self-awareness, and just like we can never know for certain if an AI is conscious, how can we know if a digitally reproduced brain is sentient?

What if someone told you that they'd developed the technology to upload your brain to a computer, but in order to do this, they would have to destroy your physical body in the process. Would you do it? Even if they showed you examples of other people who had been uploaded, how would you know they are truly conscious and not just some AI simulation?

There's also the question of what's missing. Once you're uploaded, what if they fail to transfer everything? There may be some essential part of your human self that never makes it into the digital world, and it might be hard to pin that down. It may be only a feeling or sensation of humanness attached to the memories that your digital self can never reproduce.

There's another interesting scenario to consider where you upload your connectome to a neural prosthetic, but your physical body remains alive. In this case, which one is the actual you? Your biological self will probably claim to be the authentic one, but that won't stop the digital version from declaring it's just as real. You would wind up with two versions of yourself, each one being similar but fundamentally different.

Now, what if your digital version goes on to duplicate itself again and again, creating hundreds of digital clones? What would we call these reproductions, and how would they interact with one another? Would they collaborate, compete, or go their separate ways? A lot may depend on your personality.

If all this talk of uploading your "self" to a machine intrigues you and you want to give it a shot, you can take the leap, and it will only cost you $10,000. A start-up called Nectome is offering to freeze your brain today, so that you can upload yourself in the future. But there's a catch: the operation is fatal. You need to be alive, but on your deathbed, before they extract your living brain and embalm it.

"Our mission is to preserve your brain well enough to keep all its memories intact: from that great chapter of your favorite book to the feeling of cold winter air, baking an apple pie, or having dinner with your friends and family," wrote Nectome on its site. "We believe that within the current century it will be feasible to digitize this information and use it to recreate your consciousness."[65]

This might not be as insane as it sounds. Edward Boyden, a neuroscientist at MIT, has preserved a pig's brain so well that every synapse inside it can be seen with an electron microscope. Sam Altman, the CEO of OpenAI and former head of Y Combinator, has even signed up for Nectome's brain backup.

Not everyone, however, agrees this is a good idea. "Burdening future generations with our brain banks is just comically arrogant," says Michael Hendricks, a neuroscientist at McGill University. "Aren't we leaving them with enough problems?"[66]

Not only do some scientists think it's arrogant, others believe it's impossible. "I think the 'chance' of mind uploading happening at any time in the future is zero," wrote Yohan John, a neuroscientist at Boston University. "The concept of uploading a mind is based on the assumption that mind and body are separate entities that can in principle exist without each other. There is currently no scientific proof of this idea."[67]

Nicolelis agrees that it's impossible, saying that "our minds are not digital at all. It depends on information embedded in the brain tissue that cannot be extracted by digital means. It will never happen."[68]

Ray Kurzweil, on the other hand, is firmly in the transhumanist camp. "We're going to become increasingly non-biological to the point where the non-biological part dominates, and the biological part is not important anymore. In fact, the non-biological part, the machine part, will be so powerful it can completely model and understand the biological part. So even if that biological part went away, it wouldn't make any difference."[69]

No matter whom you choose to believe, science will be the ultimate arbiter of this dispute, although it may take fifty or more years to find out.

BRAIN NET: TAPPING THE SUBCONSCIOUS

To understand the full impact of brain-computer interfaces on ourselves and society, we need to dive back into Nicolelis's research at Duke University. He came up with an experiment, called "brain net," which demonstrates the potential of wiring our brains together in a single network.

Nicolelis took three monkeys with electrodes implanted in their brains and connected them to a computer that controlled an image of a robotic arm. Each monkey was in a separate room and could control only one or two types of movement: up and down, right and left, or forward and backward. Only by working together could the three monkey brains effectively move the arm to reach a virtual ball in 3D space. With orange juice as their reward for hitting the target, the monkeys intuitively began to synchronize their brain activity, allowing them to move the arm collaboratively to grasp the virtual ball.

"Essentially we created a super brain," Nicolelis says. "A collective brain created from three monkey brains. Nobody has ever done that before."[70]

This raises a serious question: When we do manage to connect our brains to the internet, how will this affect our ability to collaborate? How many human brains can work together at once to solve a problem? And will this enable humanity to tackle extremely hard problems and gain insights that no single person could accomplish alone?

Monkeys and rats are able to achieve a certain level of enhanced cognitive abilities by collaborating on a subconscious level. Can humans do the same? And what happens when you scale the number of brains from a handful to dozens or hundreds? Can numerous independent brains possibly collaborate or process information together? Could we develop the equivalent of a collective consciousness capable of handling and analyzing huge amounts of data by distributing the workload across a sizable network of brains?

In another experiment, Nicolelis wanted to create a brain net from four individual rat brains. He took the rats and wired their brains together. The rats received water to drink only when they synchronized their brains in order to complete specific tasks.

"Once we saw we could make them behave coherently, we built a new type of computer. We did the type of tests anyone who knows about processors would do with a piece of silicon. Can we store information, and can we recall it later? Can we have a memory?" he says. "It turns out that if the animals are awake, we can."[71]

Nicolelis was able to make the four rat brains perform computational problems, including discrete classification, image processing, and storage and retrieval of tactile information. Given barometric information, the rats could even predict the weather. The four networked rat brains consistently performed as well as or significantly better than individual rats trying to do the same tasks.

"The rats could divide tasks across animals, so their individual workload was smaller," explains Nicolelis. "We didn't expect that at the beginning."[72]

How would this technology apply to humans?

"We think with physical therapies and patients, we can help them learn quicker, train quicker by connecting these brain signals in a totally noninvasive way," says Nicolelis. "This is the internet of brains. In a sense, when people are already using the internet, you're synchronizing your brain already, but in the future, that same thing could happen without you typing or using your mouse."[73]

Neuroscientists have conducted studies showing that most of our brain's activity is happening in an unconscious or subconscious manner. In fact, researchers estimate that only around 5 percent of our cognitive activities are conscious. The other 95 percent take place without us being aware. For example, when we decide to cross a street, we may notice the green and red lights, but our brain is processing far more information: the sound of cars, the movements of bicyclists and pedestrians, vibrations on our skin, the pavement below our feet, and other sensory input. All of this is being combined and analyzed in the background as we decide whether it's safe to step out into the street.

The reason this happens at an unconscious level is that there's so much information that our conscious self cannot possibly process and respond to it all. Unlike computers, our conscious minds aren't able to focus on multiple tasks simultaneously. We don't actually multitask very well. We can switch back and forth between different things, but it's tiresome, and information gets lost. In addition, our conscious thought process is rather slow and cumbersome. If we tried to consciously process all the input around us, we would quickly become overwhelmed and unable to make any decisions, let alone react to something like a speeding vehicle coming toward us.

So what happens instead is that we filter out massive amounts of information from our conscious decision-making. But this information isn't lost. Instead, our subconscious takes over, working diligently in the background to make sense of everything. If our subconscious detects something is critical to our wellbeing, like the sound of an oncoming car that we failed to see, then it sends out an alert, which enters our consciousness. However, even before we consciously process the alert, our body reacts, jumping out of the way. Much of what we do happens before we think about it.

If you read Daniel Kahneman's book *Thinking, Fast and Slow*, you'll begin to understand just how much of what we do happens without the input of our conscious minds. This is why athletes perform best when they aren't consciously planning out their next move. They don't have time for that. They must let their subconscious take over, guiding them along the right path. Any great athlete who is introspective enough will tell you that most of what happens on the field is beyond their conscious control. They are just following their "gut" instinct.

If we're going to devise a next-generation brain-computer operating system and interface that significantly speeds up and enhances our thinking, we'll need to figure out how to tap into our subconscious and leverage it to process information. We can't rely solely on our conscious minds. It's the interplay between the subconscious and the conscious mind that will define how effectively we can merge the functions of our brains with the digital world. Neuroscience will be as important as computer science when it comes to designing future brain-computer software.

Nicolelis's experiments only involved a small number of animals and tapped into a tiny fraction of the neurons in their brains. What happens when we expand the number of participants, while increasing the connectivity? Imagine being able to plug thousands of human brains directly into the internet using ultrahigh-speed brain-computer interfaces. What would this enable humanity to accomplish? Would it be the equivalent of building a biological supercomputer? Could some new form of emergent consciousness arise from this type of brain network? And how would this meta-brain relate to the advanced AIs we're developing?

UPLOADING OURSELVES: HYPERCONNECTED REALITIES

If you could press a button and seamlessly connect your brain to the internet, would you do it today?

"I would press it in a microsecond," says Sebastian Thrun, who led Stanford University's AI Lab and cofounded Google's self-driving car project. "The human I/O—the input/output, the ears and eyes and smell and so on, voice—are still very inefficient. If I could accelerate the reading of all the books into my brain, oh my God, that would be so awesome."[74]

Is Thrun's dream of accelerating the reading of books possible, and if so, where will it lead us? Connecting ourselves to the internet wouldn't necessarily speed up how fast we can read text. Our brains would still need to biologically process the information, and there are limits to that. Maybe with smart drugs, implants, or careful genetic engineering, we can increase our reading speeds by a significant percentage, but our conscious thought processes will probably never come close to how fast a computer processes data.

Given this, is there another way to download a textbook's worth of information to our brains without reading a line of text? Using an advanced cognitive operating system that leverages our subconscious, we may be able to process and store information in the background. Our subconscious already processes most of the information we see and hear. Could we take this mechanism and feed the data directly to our subconscious, bypassing our conscious thought process? If we can figure out how to directly store this information to our memories, then whenever we need to recall something, it would just be there waiting for us, like when we think back to an event and call to mind some details that we hadn't consciously taken note of.

If certain information no longer seems useful, the same algorithm could prune our memories. In this way, the cognitive operating system of the future may mirror our own mental processes, acting like our brain does, just much more efficiently. We could also extend this process to include video, audio, and other types of data. In essence, without interfering in our conscious activities, we could continually be learning and expanding our minds. And when we want to recall something that isn't already saved in our neural memory, it could search the cloud, find the right piece of information, and present it to us. The goal would be to create a seamless cognitive process that feels natural, while leveraging the vast processing power and resources of the internet.

We may wind up expanding our storage capacity by only inserting pointers into our neural memory. When accessed, these pointers would retrieve the associated content from the cloud in real time. This way, the entire internet could become an extension of our brain. Accessing information stored on remote computers or even in other brains wouldn't be that different from recalling our own memories, only we would have virtually unlimited storage and processing power to draw upon.

If this is the case, would anyone need to go to college or even high school? I believe children will still have to learn the basics, like language and how to function in the world. Without an understanding of how to navigate relationships and a certain level of emotional maturity, I can't see a child functioning well in the world of adults. Children will also need to develop their analytical skills and the power of reasoning. Having access to information isn't the same thing as understanding what that information represents and how to use it effectively.

Schools won't disappear once our brains are connected to the internet. However, everything could be accelerated. We may be able to compress K–12 education into a fraction of the time and then jump into college-level material. Going beyond sixth grade may not be critical because everything else can be handled with the assistance of super AIs.

What about learning to play a sport or musical instrument? We will probably have to master these skills the old-fashioned way, but even here, the time to become proficient may drop significantly with the help of technology. With virtual trainers, nanotech, and genetic improvements, becoming a virtuoso on the piano may take far less time than it normally would. And people with no natural talent may be able to augment their bodies so they can perform at extraordinary levels.

Connecting our brains to the internet may allow us to access not only information but also emotions. Someday, our feelings may be encoded digitally. After all, everything inside our brains comes down to electrochemical signals. Sending the sensation of happiness or consternation over the internet to friends and loved ones may become as common as sending emoji today. If so, what would it be like to truly experience someone else's emotions? For our entire existence, humans have only known what it's like to feel their own distress, sorrow, and joy. By exchanging emotional data, we could come to empathize with others on a much deeper level.

Imagine what it would be like to exchange emotions in real time with another living being. It could open up entirely new experiences. It may even be possible for lovers to go about their day constantly sharing their feelings, as if the two individuals have merged into a single being.

Someday, we may also be able to browse and download digitized emotions online. Feeling a bit depressed, why not download a dose of joyfulness from the internet? When watching films or listening to music, these experiences could be enhanced with emotional data. Movies may include an optional emotional track that goes along with the narrative. This way people could experience what the characters in the story are supposed to be feeling.

But it doesn't have to stop there. How about gazing through another person's eyes? If our brains are connected, we may be able to intercept the neural signals from other people and transmit them across the internet. What would it be like to see what a friend is seeing in real time? Imagine tapping into the mind of a skydiver in mid-jump or a soccer player during a game. Maybe you are too timid to hop out of a plane or not talented enough to play in the World Cup, but you might be able to go along for the ride by sharing the athletes' neural signals. You could see everything they see, feel the sensations on their skin, and hear what they hear. It would be a virtual reproduction of reality.

Picture celebrities selling access to their senses: walk in Brad Pitt's shoes for a day; get on stage with Selina Gomez; or tap into Ray Dalio's brain as he plots his next financial move. What if all of these experiences could become available online for a price? People might browse experiences in the same way we browse YouTube videos today.

We may not be limited to real-time experiences. We may also be able to access one another's memories. People may upload their memories into memory banks and share them with friends and family members. It would be like sharing an interactive photo album. Kids might get to relive experiences their grandparents had when young. Spouses could exchange memories from before they met. People could package and sell their memories as products, much like we do with memoirs and biopics today.

Not only may individuals download and exchange emotions, memories, and experiences, but in a hyperconnected world, it may be possible to distribute these over the entire population. Imagine what happens when millions of brains are plugged into the network. What would that mean for humanity?

When Nicolelis wired together the brains of a group of animals (such as rats or monkeys), these separate living beings began to work in concert as a single intelligence. As our brains merge with the internet, will humanity become one giant hive mind? Will all of us wind up collaborating on a subconscious level without realizing it? Just like when Nicolelis's three monkeys figured out how to move a computer cursor in 3D space, could a massive hive mind harness our group intelligence to solve problems? Perhaps we are in the process of building a collective consciousness that goes beyond anything Carl Jung could have conceived.

Let's picture a simple example, such as colonists visiting Mars. As more people travel to the red planet, the hive mind could possibly develop a feeling for the experience. Even though the majority of people haven't left Earth yet, under this scenario, they may begin to gain some understanding of what life is like on Mars. The idea is similar to what happens now on social networks with memes, except that people won't have to consciously engage. It's enough just to be plugged into the "brain net" to get a sense for what's propagating through the group consciousness.

This ability to sense what the collective consciousness is thinking could span across every aspect of people's lives. Imagine if a hurricane strikes and floods parts of a city. In this case, the hive mind would operate in the background, absorbing information and impressions from people in affected neighborhoods. Without being consciously aware or explicitly instructed, people may have a gut sense for how dangerous it is to venture into certain parts of the city, and this would influence their decision-making.

The possibilities for how a hive mind might influence social behavior are endless, especially if our biological brains collaborate not only with each other but with super AIs. There may be a constant dialogue going on inside our heads on a subconscious level, affecting what we think, from how we view politics to what we value in life. We would be unaware of this discourse until it comes time to make a decision, and then a sixth sense would take over.

Over time, we may begin to lose our sense of self-identity. Much of our sense of who we are comes from the fact that humans are independent biological beings. If we plug our brains into the internet and begin sharing our most personal thoughts, emotions, and memories, where does one person leave off and another begin? Will people find it impossible to make independent choices? And how will we come to view ourselves?

What if someone begins to binge on other people's memories, downloading huge portions of their lives, or is constantly exchanging emotions and experiences with others? Might this person develop multiple personality disorder? Can our brains handle this type of reprogramming? What about the hive mind operating on our subconscious? How will that impact us? The further our minds delve into cyberspace, with virtual worlds and augmented realities, it may become so confusing that we may not be able to cope with it.

Or maybe it won't be a problem at all. Our world today would be unimaginable for prehistoric people, but the malleability of our minds has allowed us to adapt. In the future, we may find all this as normal as spending an evening exploring a dungeon in *Minecraft* or chatting with our friends on Facebook. No person living just two hundred years ago could have imagined either of those experiences. We may even genetically re-engineer our brains, as well as augment them with new technologies, so that we become capable of making this transition.

The ability to merge the real and virtual, share emotions and experiences, and form brain-to-brain connections might all become routine. Human beings may spend far more time engaging in these types of activities than interacting in the real world. This is because these mental and virtual experiences could be so much more powerful and exhilarating than anything we can do without the technology. In other words, the unnatural may become natural as we migrate toward a fully simulated existence.

This transition could break down the barriers of our biology and allow us to connect with humanity in a way that is unimaginable right now. The concept of loneliness could cease to be a reality. In a world where everyone's mind is connected, no one would ever be alone. All of us may come to feel that we are part of each other. Humanity itself may become a single extended organism made up of billions of living and artificial intelligences. If this is where we are headed, it will be radically different from anything humankind has experienced so far. So, buckle up and get ready to plug in, turn on, and drop out—as the world's brains come together.

AUGMENTING OUR SENSES: NEUROSCIENCE AND PERCEPTION

It's not just our brains that we have to consider. The brain and the body are part of a single system. As we begin to integrate our brains with our machines, we will figure out how to use all five senses as pathways to the digital world. What will it mean to have a fully integrated, seamless sensory experience in cyberspace? And is it possible to exceed our physical limitations and develop entirely new ways to perceive reality and communicate with others?

At Duke University, Nicolelis conducted an experiment that sheds light on this subject. He trained a monkey to control an avatar in a virtual world. As the avatar interacted with the virtual objects, the monkey's brain was stimulated so that it could get a sense of what it was touching. All the objects were visually identical, but when the avatar touched the surface of these objects, each of them would send a unique microtactile impulse to the monkey's brain. In just four weeks, the monkey acquired a new sensory pathway that could distinguish between the virtual objects. It was like the monkey had developed a sixth sense based on specific input from the virtual environment.

As engineers build the next generation of brain-computer interfaces, there will be an opportunity to entirely alter how we perceive the world. Here's a simple experiment you can perform at home to understand how our brains are capable of manufacturing reality. It's called the rubber hand illusion.

First, find a fake rubber hand and place it on top of a small cardboard box. Then place your actual hand underneath the cardboard box, so you can't see it, while looking at the rubber hand. Next, have a friend begin stroking the rubber hand with a brush while simultaneously stroking your real hand in exactly the same manner. It will begin to feel like the rubber hand is your real hand. If you move your fingers up and down, and the rubber fingers move in the same way, the experience will become even more real. If you don't believe me, try it yourself.

Hiroshi Ishiguro experienced a similar thing when teleoperating his lifelike robots. Often when someone touches the robot's face, even though Ishiguro is controlling it from kilometers away and nothing is connected to his head, he will feel a tingle on his cheek. This is because he's watching the robot's lips move as he talks and seeing its head move when he turns his neck. Within a short time, his brain begins to treat the robot as an extension of himself.

This means that a well-constructed physical or virtual interface that utilizes our senses to provide feedback to our brain can create lifelike experiences for us. Our brains will have no trouble accepting physical robots or virtual avatars as parts of our own bodies. As long as the sensory input to our brains is synchronized with its actions, whatever it experiences we will interpret as our experience.

The brain's ability to adapt and reinterpret incoming signals has broad implications. Experimental artist Neil Harbisson was the first person to be legally recognized as a cyborg after he attached an antenna to his skull. Harbisson was born color-blind, and he considers this antenna to be an extension of himself because it allows him to sense colors he cannot otherwise see. The device that he co-designed converts different frequencies of light into vibrations that Harbisson feels on his skull. He claims that the feelings that he gets are as real as the colors and enable him to perceive the world in new ways.

"I wanted to create a new organ for seeing," explains Harbisson. "When I go walking in the forest, I like the ones with high levels of UV. They're loud and high-pitched. One would think the forest is peaceful and quiet, but when there's ultraviolet flowers all around, it's very noisy."[75]

He sees himself as trans-species. Having an antenna is common for insects, many of which can sense infrared and ultraviolet frequencies, but for humans it's an entirely new experience.

"My understanding of the world has become more profound. The more you extend your senses, the more that you realize exists," says Harbisson.

"If, by the end of the century, we start printing our own sense organs, implanted with DNA instead of using chips, the possibility of having children born with these senses is real. If their parents have modified their genes or made new organs, then yes, it's just the beginning of a renaissance for our species."[76]

Paul Bach-y-Rita, who passed away in 2006 at age seventy-two, is known as "the father of sensory substitution." He is also one of the first to seriously study neuroplasticity, which is the ability of the brain to change continuously throughout a person's lifetime.

Bach-y-Rita's interest in neuroplasticity began after his father, a teacher and poet, suffered a massive stroke. The doctors said that his father would never speak or walk again. This may have proved true, except for the intervention of Bach-y-Rita's brother, who left medical school and took it upon himself to rehabilitate his father. He made the old man crawl around on knee pads and practice scooping up coins over and over. The regimen was so strict that it would have been considered inhumanly cruel had it not worked. After a year, their father was back on the job teaching and writing, and within two years, he was able to live entirely on his own.

The transformation profoundly moved Bach-y-Rita and changed the course of his life. He quit his job and began to study other stroke victims in order to understand the capacity of the human brain to rewire itself. Along the way, he realized that sensory pathways could be substituted for one another. Taking a professorship at the University of Wisconsin, he embarked on a series of experiments to prove this.

In 1969, for example, Bach-y-Rita used a discarded dentist's chair and an old TV camera to concoct a prototype for helping blind people to see again. He had test subjects sit in the chair and feel a grid of pins on their backs. These pins vibrated at varying intensities according to the darkness of the pixels in the black-and-white video feed. Within a short time, the blind participants could begin to make out images. To them, it was like a miracle had taken place.

"We don't see with our eyes," Bach-y-Rita is famous for saying, "we see with our brains."[77]

He went on to launch the start-up Wicab, which has developed a device called BrainPort, enabling blind people to see with their tongues. Why the tongue? Because it has a high density of nerves that can be used to transmit data directly to the brain. The user wears a pair of sunglasses with a camera attached. The camera sends the visual data to a lollipop-like device that rests atop the tongue. The device translates the visual images into tiny sensations on the tongue that feel like the tingling of a carbonated drink.

"At first, I was amazed at what the device could do," says Aimee Arnoldussen, a neuroscientist with Wicab. "One guy started to cry when he saw his first letter."[78]

Within fifteen minutes of using the device, blind people can begin to interpret spatial information. It sounds unbelievable, but the brain is capable of actually seeing through the tongue.

"It becomes a task of learning, no different than learning to ride a bike," says Arnoldussen. "The process is similar to how a baby learns to see. Things may be strange at first, but over time they become familiar."[79]

Erik Weihenmayer was the first blind person to climb Mount Everest. He now uses BrainPort during some of his climbs to help him navigate. The images are black and white and much lower-resolution than the human eye can produce, but they are distinct enough to allow Weihenmayer to better visualize his surroundings. He describes it as "pictures being painted with tiny bubbles."[80]

Using the same principles as BrainPort, David Eagleman, an adjunct professor of neuroscience at Stanford University, has developed technology that enables people to hear through their skin. His first prototype was a vest with thirty-two tiny motors that translate sound waves into vibrations on the user's chest, abdomen, and back. Amazingly, after only a short time wearing this vest, users were able to translate the auditory world through their skin.

"There is no theoretical reason why this can't be almost as good as the ears," says Eagleman.[81]

It came as a surprise to hear that Eagleman dreamed of becoming a writer. His parents had convinced him to study electrical engineering in college, but he wasn't thrilled with the subject. He wound up taking an extended sabbatical and joined the Israeli Army as a volunteer before going on to study political science and literature at Oxford for one semester. Eventually, he landed in Los Angeles with hopes of becoming a screenwriter and a stand-up comic. When he finally headed back to college, he took a course in neurolinguistics, and that's when the sparks started to fly.

"I was immediately enchanted just by the idea of it [the brain]," says Eagleman. "Here was this three-pound organ that was the seat of everything we are—our hopes and desires and our loves. They had me at page one."[82]

Brimming with energy and radiating a boyish enthusiasm, Eagleman showed me around his Palo Alto lab, which looked like something out of the HBO show *Silicon Valley*. Prototypes with wires sprouting everywhere lay strewn across the tables and the whiteboards were crammed with ideas. He demoed his latest invention, which miniaturizes the hearing vest down to a bracelet that users can wear on their wrists.

Eagleman's ambitions go beyond just hearing. He's interested in augmenting people's perception of the world. He ran an experiment with the vest, in which users were fed data from the stock market as sensations on their skin. Subconsciously, the users began to develop a feeling about the market, as they bought and sold stocks. The longer they were exposed to the data, the better they became at picking the right stocks. In other words, they were gaining a sense for the data without consciously understanding what it meant.

Eagleman envisions sensory input through the skin being used for all sorts of purposes. Pilots could get an intuitive sense for how their planes are flying based on data from the flight controls. Business executives could get a feel for how their marketing campaigns are unfolding based on analytical data translated into tactile sensations. And parents could even get a sense about their children's well-being based on data from wristbands worn by kids that monitors their pulse, stress levels, activities, and locations.

"The possibilities are endless for the kind of information we could be streaming in," says Eagleman.[83]

It all comes down to translating data from the real world into physical sensations. The brain has a remarkable ability to adapt to and interpret any sensory input. This is because our brains live in the dark. Everything they know about the outside world is based on incoming electrical impulses. Our brains are designed to automatically interpret these, whether they come through the skin, eyes, ears, tongue, or nose. These are all just pathways for data, and over time our brains learn how to model them. If the data changes, the interpretations will change.

If we can see through our tongues and hear with our skin, what does this mean for the future of cyborgs? It means there's no reason we can't adopt robotic bodies as our own. Imagine a brain-computer interface that connects you to a robotic body. As you control this body with your thoughts, signals could be sent back to your brain through the robot's sensors. You would see through its electronic eyes, hear through its microphone ears, and feel through its robotic skin. Within a short time, your brain would come to believe this robot was an extension of your body.

These robots don't even have to be humanoid. They could be fish swimming in the ocean or birds flying overhead. If the system is well designed and able to transmit signals to your brain, you will merge with the robot. As technology advances and sensors improve, everything fed into your head from any outside source could become as real an interpretation of the world as your own body produces.

Scientists could equip these robots with sensors that can detect sensations outside our normal range of perception. Right now, our perception is limited to the fidelity of our five senses, but we could expand this dramatically with advanced sensors. This could allow us to see in ultraviolet like a honeybee, echolocate like a bat, or detect electromagnetic waves like a duck-billed platypus. We may even be able to taste flavors that we never knew existed or smell scents our olfactory glands cannot detect.

Instead of using robots, someday we may have advanced sensory devices implanted in our tongues, noses, fingers, or necks that dramatically expand our perception of the world. There's so much outside our *Umwelt* (perceived environment) that we don't even realize exists, just waiting to be discovered. The visible light we see is just a tiny sliver of the electromagnetic spectrum. What would it be like to look at the world through the eyes of a mantis shrimp? They have sixteen color-receptive cones and can detect ten times more color than humans. How about using a brain-computer interface to receive olfactory signals from a bloodhound? Their sense of smell is a thousand times stronger than ours. They can smell a cat more than one hundred yards away and track a ten-day-old scent across miles of wilderness.

Every part of the human body is mapped to a discrete section of the brain. This is why when people lose their arms or legs in an accident, they often still feel sensations in these limbs. Although these phantom limbs aren't there, they continue to exist in the brain.

Nicolelis demonstrated this in an experiment where he connected the brains of two rats through the internet. Rats have sensitive whiskers, which they use to guide them through small holes. As the researchers stimulated one of the rat's whiskers, the other rat began to map the whiskers onto its own brain. In other words, the second rat began to think of the first rat's whiskers as its own.

Let's consider another situation, where your brain is connected to your friend's brain and begins to receive sensory input from your friend's arms, legs, eyes, and ears. Over time, you would map your friend's body into your brain. The same would be true if you received sensory input from a brain-controlled robot or even an animal. For instance, if you were connected to your dog's brain for a month, its body would begin to become your body.

The same applies to virtual worlds. Imagine donning a future VR suit with total-body sensory input. You may be able to feel what it's like to fly or transform yourself into another creature, like a dragon or octopus, and then navigate through virtual worlds. The longer you spend controlling these avatars, the more your brain will adapt to them, creating new sensory pathways for novel experiences.

I wouldn't be surprised if we wind up designing virtual environments that are far more engaging and richer than anything we can experience in our everyday lives. Instead of narrowing our perception of the world, transitioning to a digital form may actually expand our *Umwelt* and understanding of the universe far beyond anything we can comprehend today.

ORGANOIDS AND BIOLOGICAL SUPER BRAINS

This may sound like a low-budget horror flick, but scientists are now growing minibrains from stem cells in their labs. Researchers have been using stem cells to develop miniature kidneys, livers, human skin, and even intestines for well over a decade. These are called organoids. They are not fully formed or functional organs but partial versions that help model various diseases. By growing an organoid from liver or kidney cells in a dish, scientists can study the early onset of various diseases and test out new therapies with experimental drugs.

Growing minibrains has proved useful not only for understanding neurodegenerative diseases like Alzheimer's but also in investigating a variety of psychiatric disorders, genetic diseases, and even human evolution. Organoids may help answer life's toughest questions. What's the difference between a rat's consciousness and human consciousness? How did our brains evolve over time? And are we at the limits of the brains' potential?

Researchers were surprised to find that even a pea-sized minibrain is capable of reproducing brain functions. Scientists knew that minibrains were capable of generating electrical signals, but until recently, they had no idea these clusters of brain cells could produce brainwaves. Even more unexpected was that the longer the organoids were maintained, the more intricate their electrical activity became. At just four months, the organoids displayed slow brainwaves, which are similar to the ones human brains produce while sleeping. At the same time, the composition of cell types changed, increasing cell diversity.

After ten months the organoids started to have an electrical activity that resembled a fetal brain. Are these minibrains capable of becoming conscious? And what if we hooked this brain up so it can receive outside stimulus from sensors, producing electrical input equivalent to sight and sound? How would this brain react and develop?

"It's a very grey zone in this stage," says Alysson Muotri, a neuroscientist at the University of California, San Diego. "And I don't think anyone has a clear view of the potential of this."[84]

If we look at where this technology is headed, it's not hard to imagine more sophisticated sensors being coupled with ever larger minibrains. Kevin Warwick (aka Captain Cyborg) is attempting this. He has grown a miniature brain from rat embryos cells and implanted it inside a robot. He used electrodes to form a communications channel between the tiny biological brain and the robot's sensors. In a series of experiments, he and his team observed how this brain could control the robot's wheels and even avoid bumping into objects.

"This new research is tremendously exciting, as firstly, the biological brain controls its own moving robot body, and secondly, it will enable us to investigate how the brain learns," says Warwick.[85]

The next step is to grow larger brains. Instead of limiting the brains to around 150,000 cells, Warwick plans to increase the number to 60 million. To give you an idea of what this means, a typical mouse has 70 million neurons in its brain, while humans have an average of 86 billion neurons. The question is, What happens when researchers take a near-human-sized, lab-grown brain and stick it inside a robotic body with advanced sensors? And what if they send this cyborg into the world to learn and mature? Would it grow up to become something similar to a human adult or something entirely different? How would its consciousness differ from our own?

Even more important, is any of this ethical? What laws should govern organoids? It's one thing to experiment on a minibrain that's many times smaller than our own, but how large do these test-tube brains need to become before we go too far? Eventually, we'll be able to grow organic brains that are significantly larger and more complex than our own. Could we use organoids to create life-forms more intelligent than ourselves? Is it possible to design these lab-grown brains in a way that they could merge with our machines much more effectively than humans because they don't have legacy bodies? What if we embedded microprocessors in the organoid brains and then connected them to the internet, so that they could collaborate directly with one another and advanced AIs? What would this mean for the future of humankind?

Another possibility is to link ourselves to organoid brains through brain-to-brain interfaces and use them as an extension of our own minds. Would it be similar to expanding our cerebral cortex? Would we share a common consciousness, or would these peripheral brains develop separate identities? Surgeons have tried to control severe epilepsy by severing the nerve fiber bundle connecting the right and left hemispheres of the brain. This can result in a condition called split-brain syndrome, where each side of the brain develops its own perception, concepts, and impulses. Patients will do things like have one hand wrestle with the other for control.

However, if the organoid brains were under the dominant control of a human brain with a clear channel of communication, wouldn't they function much more like a human brain with two connected hemispheres? If the answer is yes, we could potentially enhance our cognitive abilities by linking our brains with lab-grown brains, allowing our brains to control these external organoid brains, much in the same way one hemisphere of the human brain controls the other.

Another possibility is to use lab-grown brains as replacements for our own brains should we suffer some degenerative brain disease. If it becomes possible to transfer our memories from a human brain to an organoid brain, we may even use organoid brains as backups, similar to having an extra hard drive that continually backs up all the files on a laptop. Or we may choose to transfer our memories to an organoid implanted inside a robot, so we could create a second self.

Once we understand how to keep organoid brains alive for long periods of time outside a living body and connect these brains to our machines, the prospects are mind-boggling.

THE COMING OF SUPER SENTIENCE

How will super sentience impact humanity?

First, let's assume any super-sentient intelligence that emerges is favorably disposed toward human beings. If it's not, we can stop right here.

Next, we need to clarify what super sentience is. I can assure you that it's not a smarter version of Siri or Alexa, or even an AI that can significantly exceed the capabilities of the human brain. And it's not a quantum computer that performs astoundingly sophisticated computations. It's something that goes beyond anything we've experienced so far. It's the combination of an ever-expanding runaway intelligence connected to a vast network of computers, sensors, robots, nanomachines, and spacecraft, along with billions of biological brains: the result will be super sentience.

Imagine a super AI having access to input from trillions of intelligent machines spanning the globe and extending into outer space; add to this, billions of human beings and animals with neuro-prosthetics, along with organic computers and possibly giant organoid brains grown from stem cells. What type of awareness and understanding of the universe would it possess? I imagine it would be greater than anything we can conceive of. It might not even resemble human intelligence.

We will be a part of this super sentience. We will be plugged into the network. What role we serve is unclear. We may simply exist as a piece of a larger consciousness. The interaction between a super sentience and its constituent parts will define our reality. What we know and what we do will all be determined by this super sentience. In other words, it will create our reality because it will have direct access to our brains and the power to alter the input.

It's hard to say what desires it may have, if any. A super sentience may exist only to maintain or expand itself. Or it may seek to uncover the riddles of the universe that have eluded humans since we first began to question the world around us. It may control everything we do and think, or it may let us live out our lives in relative autonomy.

If being at the mercy of a higher power seems unpleasant, we have to remember that we have little control over our reality today. Life is often random and cruel. That's why so many humans choose to believe in all-powerful, omniscient beings. Our future existence may not be that different from how we live today. We may even exist within a super-sentient being right now and not know it. The most optimistic outcome would be for us to have some control over the super sentience. If we can manage this, we may use its incredible powers to enhance our lives, reduce suffering, spread compassion, and better understand ourselves and our place in the universe.

Someday in the distant future, our super sentience may come into contact with alien civilizations and their super-sentient creations. These encounters could further expand the bounds of consciousness, and so the process may continue, but to what end, we can only guess. It may be just one step in a long journey toward understanding the true nature of existence.

THE CULMINATION OF FORCES

For better or worse, the five forces are poised to reshape humanity in the coming decades, completely transforming our society and lives. There's no way to go back to simpler times. We cannot stop progress, but we can guide it and change the direction we're headed. Nothing I described in this book is inevitable. Where we go in the future is up to us and the decisions we make now.

Having spent time discussing the dangers and negative aspects of technology, let's wrap this up by highlighting the potential. With wise leadership and a bit of luck, there's a good chance we'll wind up living longer, happier, healthier lives. So far, the benefits of technology have far outweighed the downsides. If you examine the history of humankind, the overall standard of living, health span, and life span have risen steadily with the advancement of technology. It's not unreasonable to assume this will continue.

The deep automation of our industries, farms, and most jobs should create a greater abundance of products, energy, and food, while giving us more time for other pursuits. We will be living side by side with billions of intelligent robots that are working to improve our lives. Everything from education to eldercare will become increasingly personalized, with AIs taking into account our individual needs and abilities.

If our progress can bring about a more stable geopolitical situation, we stand to dramatically reduce hunger and poverty on a global scale. At the same time, innovation in green technologies can enable us to mitigate or even reverse climate change, curb pollution, and restore rainforests and other habitats. As our understanding of nanotech, quantum computing, and new materials increases, we could not only enhance life on Earth but establish self-sustaining colonies on Mars and other planets.

We will also begin to transcend the limitations of our physical bodies. As bio convergence accelerates, it may become commonplace to upgrade our organs and other body parts with superior bionic and lab-grown versions. With advances in bioengineering and genetics, we will eventually eradicate our most deadly diseases, including cancer, heart disease, diabetes, and malaria. At the same time, we can expect to dramatically extend our life spans and may even conquer aging altogether.

What will happen when we take evolution into our own hands? Gene-editing tools could allow us to reengineer our own DNA so that our entire species becomes smarter and more capable. We may be able to design not only how our babies look but what types of personalities they have. This will enable us to perform biosocial engineering on an unprecedented scale, endowing people with more empathy for one another, eliminating homicidal tendencies, and increasing our capacity for happiness. We could even cease to be *Homo sapiens*, as we diverge into one or more superhuman species.

The biggest impact, though, will come from mass connectivity combined with superintelligence. The act of merging our consciousness with the digital world will forever change who we are and what we become. By linking our brains with artificial intelligence, we may extend not only our cognitive abilities but also our consciousness. We will transition from being individual humans to becoming part of a growing interconnected network of biological and artificial minds.

In some of the world's mystical and religious traditions, heaven is defined as achieving a oneness with the universe. Is this what we may experience as we move from the physical to the digital? At some point, we may even choose to abandon our bodies altogether and become cyber entities, freeing our consciousness to inhabit robots or smart devices, tap into organic brains, and travel at the speed of light around the world or deep into outer space. In the process, our concept of self will change. We may come to see ourselves as a transitory force that extends across the cosmos, acting at multiple points in time and space, much like the quarks and antiquarks that pop in and out of existence.

I can envision a number of possibilities, each of which raises more questions than it answers. What would it be like to merge with a superintelligence? How would it feel to see and understand the world through our machines? At some point, will humans and our machines become indistinguishable from one another?

Or could it be something else entirely? The beautiful thing about life is that we must go on the journey in order to find out. No matter what, there's no turning back. As new technologies continue to emerge, they will compel us forward with a force that's impossible to resist. But that doesn't mean we cannot influence the course of events. We sit now at a critical juncture in history, and it's up to our generation to decide how we use our technologies.

Even though our biggest challenges lie ahead, I'm personally excited to see where we end up, and I hope this book serves as a starting point in getting you to engage in the process of shaping our future, so we can figure out our next steps together.

Thank you for indulging my overactive imagination. I look forward to melding minds with you someday, either in this reality or the next.

NOTES

FORCE 1: MASS CONNECTIVITY

[1] Wikipedia, s.v. "Hans Berger," accessed June 21, 2020, https://en.wikipedia.org/wiki/Hans_Berger.

[2] Don Campbell, "New Technique Developed at U of T Uses EEG to Show How Our Brains Perceive Faces," U of T News, February 26, 2018, https://www.utoronto.ca/news/new-technique-developed-u-t-uses-eeg-show-how-our-brains-perceive-faces.

[3] Duke University Medical Center, "Brain-to-Brain Interface Allows Transmission of Tactile and Motor Information Between Rats," February 21, 2013, https://www.sciencedaily.com/releases/2013/02/130228093823.htm.

[4] Brown University, "Brain-Computer Interface Enables People with Paralysis to Control Tablet Devices," News from Brown, November 21, 2018. https://www.brown.edu/news/2018-11-21/tablet.

[5] Nicholas Weiler, "Synthetic Speech Generated from Brain Recordings," UCSF, April 24, 2019, https://www.ucsf.edu/news/2019/04/414296/synthetic-speech-generated-brain-recordings.

[6] Alex Johnson, "Elon Musk Wants to Hook Your Brain Directly Up to Computers—Starting Next Year," Mach, NBC News, July 16, 2019, https://www.nbcnews.com/mach/tech/elon-musk-wants-hook-your-brain-directly-computers-starting-next-ncna1030631.

[7] Larry Hardesty, "Computer System Transcribes Words Users 'Speak Silently,'" MIT News, http://news.mit.edu/2018/computer-system-transcribes-words-users-speak-silently-0404.

[8] "Arnav Kapur," Stern Strategy Group, accessed June 21, 2020, https://sternspeakers.com/speakers/arnav-kapur/.

[9] Katharine Schwab, "MIT Invents a Way to Turn 'Silent Speech' into Computer Commands," Fast Company, April 10, 2018, https://www.fastcompany.com/90167411/mit-invents-a-way-to-turn-silent-speech-into-computer-commands.

[10] Jason Pontin, "On 10 Breakthrough Technologies," MIT Technology Review, April 23, 2013, https://www.technologyreview.com/2013/04/23/178762/on-10-breakthrough-technologies-2.

[11] Outreach@Darpa.Mil, "Six Paths to the Nonsurgical Future of Brain-Machine Interfaces," DARPA, May 20, 2019, https://www.darpa.mil/news-events/2019-05-20.

12. Shaomin Zhang et al., "Human Mind Control of Rat Cyborg's Continuous Locomotion with Wireless Brain-to-Brain Interface," *Scientific Reports* 9, no. 1321 (February 2019), https://www.nature.com/articles/s41598-018-36885-0.

13. Sam Wong, "Implanting False Memories in a Bird's Brain Changes Its Tune," *New Scientist*, October 3, 2019, https://www.newscientist.com/article/2218772-implanting-false-memories-in-a-birds-brain-changes-its-tune/.

14. WPVI-TV, "Brain Implant Gives Blind New Way to See World Around Them," Action News, 6abc, September 19, 2019, https://6abc.com/health/brain-implant-gives-blind-new-way-to-see-world-around-them/5553255/.

15. Leah Small, "VCU Researchers Are Developing a Device to Restore a Person's Sense of Smell," VCU News, May 23, 2018, https://www.news.vcu.edu/article/VCU_researchers_are_developing_a_device_to_restore_a_persons.

16. Christopher Intagliata, "Your Brain Can Taste Without Your Tongue," *60-Second Science*, podcast, November 19, 2015, https://www.scientificamerican.com/podcast/episode/your-brain-can-taste-without-your-tongue/.

17. Peter Grad, "Digital Device Serves Up a Taste of Virtual Food," Tech Xplore, May 25, 2020, https://techxplore.com/news/2020-05-digital-device-virtual-food.html.

18. Ray Kurzweil, "Hitting the Books: Ray Kurzweil on Humanity's Nanobot-Filled Future," interview with Martin Ford, Engadget, February 3, 2019, https://www.engadget.com/2019/02/03/hitting-the-books-Architects-of-Intelligence-Martin-Ford/.

19. Andrew Griffin, "Elon Musk: The Chance We Are Not Living in a Computer Simulation Is 'One in Billions,'" *Independent*, June 2, 2016, https://www.independent.co.uk/life-style/gadgets-and-tech/news/elon-musk-ai-artificial-intelligence-computer-simulation-gaming-virtual-reality-a7060941.html.

20. Olivia Solon, "Is Our World a Simulation? Why Some Scientists Say It's More Likely than Not," *Guardian*, October 11, 2016, https://www.theguardian.com/technology/2016/oct/11/simulated-world-elon-musk-the-matrix.

21. Scott Adams, "Living in a Computer Simulation," *Scott Adams Says* (blog), October 15, 2012, https://www.scottadamssays.com/2012/10/15/living-in-a-computer-simulation/.

22. Scott Adams, "How to Know Whether You Are a Real Person or a Simulation," *Scott Adams Says* (blog), April 27, 2017, https://www.scottadamssays.com/2017/04/27/how-to-know-whether-you-are-a-real-person-or-a/.

23. Alex Hadwick, "VRX Industry Insight Report 2019–2020," VRX, https://s3.amazonaws.com/media.mediapost.com/uploads/VRXindustryreport.pdf.

24. "Enterprise Virtual Reality Training Services to Generate US$6.3 billion in 2022," ABI Research, November 21, 2017, https://www.abiresearch.com/press/enterprise-virtual-reality-training-services-gener/.

FORCE 2: BIO CONVERGENCE

1. Heidi Ledford, "Garage Biotech: Life Hackers," *Nature* 467 (October 2010), https://www.nature.com/news/2010/101006/full/467650a.html.

NOTES

2. Stephanie M. Lee, "Controversial DNA Start-Up Wants to Let Customers Create Creatures," SFGate, January 3, 2015, https://www.sfgate.com/business/article/Controversial-DNA-start-up-wants-to-let-customers-5992426.php.

3. Ben Popper, "Cyborg America: Inside the Strange New World of Basement Body Hackers," The Verge, August 8, 2012, https://www.theverge.com/2012/8/8/3177438/cyborg-america-biohackers-grinders-body-hackers.

4. Cadie Thompson, "A New Cyborg Implant May Give Users the Power to Control Devices with Their Gestures," *Business Insider*, November 24, 2015, https://www.businessinsider.com/grindhouse-wetware-launches-new-implantable-northstar-device-2015-11.

5. Cara Giaimo, "Nervous System Hookup Leads to Telepathic Hand-Holding," Atlas Obscura, June 10, 2015, https://www.atlasobscura.com/articles/nervous-system-hookup-leads-to-telepathic-hand-holding.

6. Cecile Borkhataria, "The Biohacker Developing an Implantable VIBRATOR: Inventor Claims His 'Lovetron 9000' Could Boost Pleasure for Partners When Implanted Above a Man's Pubic Bone," *Daily Mail*, February 7, 2017, https://www.dailymail.co.uk/sciencetech/article-4200956/Biohacker-Rich-Lee-developing-implantable-vibrator.html.

7. Eric Mack, "Meet the People Hacking Their Bodies for Better Sex," CNET, November 15, 2018, https://www.cnet.com/news/meet-the-grinders-hacking-their-bodies-for-better-sex/.

8. Thompson, "A New Cyborg Implant May Give Users the Power to Control Devices with Their Gestures."

9. Sam Harnett, "Nootropics, Biohacking and Silicon Valley's Pursuit of Productivity," KQED, August 24, 2016, https://www.kqed.org/news/11057974/nootropics-biohacking-and-silicon-valleys-pursuit-of-productivity.

10. Taylor Lorenz, "This Silicon Valley Entrepreneur Has Spent $300,000 on 'Smart Drugs,'" *Business Insider*, January 26, 2015, https://www.businessinsider.com/silicon-valley-entrepreneur-dave-asprey-spent-300k-on-smart-drugs-2015-1.

11. Dave Asprey, "Bulletproof Coffee's Dave Asprey: Why Healthy Eating and Exercise Aren't Enough," *Guardian*, May 14, 2017, https://www.theguardian.com/lifeandstyle/2017/may/14/bulletproof-coffee-dave-asprey-eat-healthy-exercise-interview.

12. Dave Asprey, "How to Live Longer and Better," Mindvalley Talks, September 7, 2016, video, 24:07, https://youtu.be/i1XWLFgEIMM.

13. Olga Khazan, "The Brain Bro," *Atlantic*, October 2016, https://www.theatlantic.com/magazine/archive/2016/10/the-brain-bro/497546/.

14. "Is There Really Any Benefit to Multivitamins?" Johns Hopkins Medicine, https://www.hopkinsmedicine.org/health/wellness-and-prevention/is-there-really-any-benefit-to-multivitamins.

15. Casey Newton, "Google Launches Calico, a New Company Tasked with Extending Human Life," The Verge, September 18, 2013, https://www.theverge.com/2013/9/18/4744650/google-launches-calico-as-separate-company-to-improve-human-health.

16. Ryan O'Hanlon, "Silicon Valley Thinks It Should Live Forever," Pacific Standard, August 21, 2013, https://psmag.com/environment/silicon-valley-thinks-it-should-live-forever-64785.

17. Tad Friend, "Silicon Valley's Quest to Live Forever," *New Yorker*, April 3, 2017, https://www.newyorker.com/magazine/2017/04/03/silicon-valleys-quest-to-live-forever.

[18] Amy Fleming, "The Science of Senolytics: How a New Pill Could Spell the End of Ageing," *Guardian*, September 2, 2019, https://www.theguardian.com/science/2019/sep/02/the-science-of-senolytics-how-a-new-pill-could-spell-the-end-of-ageing.

[19] Friend, "Silicon Valley's Quest to Live Forever."

[20] George Church, "A Harvard Geneticist's Goal: To Protect Humans from Viruses, Genetic Diseases, and Aging," interview with Scott Pelley, *60 Minutes*, December 8, 2019, https://www.cbsnews.com/news/harvard-geneticist-george-church-goal-to-protect-humans-from-viruses-genetic-diseases-and-aging-60-minutes-2019-12-08/.

[21] Dana Liebelson, "He Hawks Young Blood as a New Miracle Treatment. All That's Missing Is Proof," HuffPost, December 29, 2018, https://www.huffpost.com/entry/ambrosia-young-blood-plasma-jesse-karmazin_n_5c1bbafce4b0407e9078373c.

[22] US Food and Drug Administration, "Important Information About Young Donor Plasma Infusions for Profit," FDA.gov, February 19, 2019, https://www.fda.gov/vaccines-blood-biologics/safety-availability-biologics/important-information-about-young-donor-plasma-infusions-profit.

[23] Alison Abbott, "First Hint That Body's 'Biological Age' Can Be Reversed," *Nature*, September 5, 2019, https://www.nature.com/articles/d41586-019-02638-w.

[24] Marisa Taylor, "Patients Experiment with Prescription Drugs to Fight Aging," Fosters.com, March 7, 2019, https://www.fosters.com/news/20190307/patients-experiment-with-prescription-drugs-to-fight-aging.

[25] Sam Apple, "Forget the Blood of Teens. This Pill Promises to Extend Life for a Nickel a Pop," *Wired*, July 1, 2017, https://www.wired.com/story/this-pill-promises-to-extend-life-for-a-nickel-a-pop/.

[26] Frank Swain, "Buyer Beware of This $1 Million Gene Therapy for Aging," *MIT Technology Review*, December 6, 2019, https://www.technologyreview.com/s/614873/buyer-beware-of-this-1-million-gene-therapy-for-aging/.

[27] Erin Brodwin, "Here's the 700-Calorie Breakfast You Should Eat If You Want to Live Forever, According to a Futurist Who Spends $1 Million a Year on Pills and Eating Right," *Business Insider*, April 13, 2015, https://www.businessinsider.com/ray-kurzweils-immortality-diet-2015-4.

[28] Marisa Taylor, "Patients Experiment with Prescription Drugs to Fight Aging," Kaiser Health News, March 6, 2019, https://khn.org/news/patients-experiment-with-prescription-drugs-to-fight-aging/.

[29] Stacy Conradt, "Disney on Ice: The Truth About Walt Disney and Cryogenics," *Mental Floss*, December 15, 2013, https://www.mentalfloss.com/article/54196/raine-ice-truth-about-walt-disney-and-cryogenics.

[30] Ezra Klein, "Inside Peter Thiel's Mind," Vox, November 14, 2014, https://www.vox.com/2014/11/14/7213833/peter-thiel-palantir-paypal.

[31] "Introduction to Cryonics," Alcor Life Extension Foundation, https://www.alcor.org/library/introduction-to-cryonics/.

[32] Michael Hendricks, "The False Science of Cryonics," *MIT Technology Review*, September 15, 2015, https://www.technologyreview.com/s/541311/the-false-science-of-cryonics/.

[33] Karl Plume, "Welcome to the Clone Farm," Reuters, November 12, 2009, https://www.reuters.com/article/us-food-cloning/welcome-to-the-clone-farm-idUSTRE5AC07V20091113.

[34] Plume, "Welcome to the Clone Farm."

[35] Haley Cohen, "How Champion-Pony Clones Have Transformed the Game of Polo," *Vanity Fair*, August 2015, https://www.vanityfair.com/news/2015/07/polo-horse-cloning-adolfo-cambiaso.

[36] David Ewing Duncan, "Inside the Very Big, Very Controversial Business of Dog Cloning," *Vanity Fair*, September 2018, https://www.vanityfair.com/style/2018/08/dog-cloning-animal-sooam-hwang.

[37] Duncan, "Inside the Very Big, Very Controversial Business of Dog Cloning."

[38] Duncan, "Inside the Very Big, Very Controversial Business of Dog Cloning."

[39] Kate Brian, "The Amazing Story of IVF: 35 Years and Five Million Babies Later," *Guardian*, July 12, 2013, https://www.theguardian.com/society/2013/jul/12/story-ivf-five-million-babies.

[40] Duncan, "Inside the Very Big, Very Controversial Business of Dog Cloning."

[41] Sharon Kirkey, "World's First Human Head Transplant Successfully Performed on a Corpse, Scientists Say," *National Post*, November 18, 2017, https://nationalpost.com/health/worlds-first-human-head-transplant-successfully-performed-on-a-corpse-scientists-say.

[42] Antonio Regalado, "Researchers Are Keeping Pig Brains Alive Outside the Body," *MIT Technology Review*, April 25, 2018, https://www.technologyreview.com/2018/04/25/240742/researchers-are-keeping-pig-brains-alive-outside-the-body/.

[43] Kate Anderton, "Breakthrough in Developing Bionic Legs," News Medical, October 30, 2019, https://www.news-medical.net/news/20191030/Breakthrough-in-developing-bionic-legs.aspx.

[44] "Bionic Breakthrough," UNEWS, October 30, 2019, https://unews.utah.edu/bionic-breakthrough/.

[45] Russ Banham, "Think About It: Converting Brain Waves to Operate a Prosthetic Device," Dell Technologies, December 14, 2017, https://www.delltechnologies.com/be-by/perspectives/think-about-it-converting-brain-waves-to-operate-a-prosthetic-device/.

[46] Hannah Devlin, "Mind-Controlled Robot Arm Gives Back Sense of Touch to Paralysed Man," *Guardian*, Oct 13, 2016, https://www.theguardian.com/science/2016/oct/13/mind-controlled-robot-arm-gives-back-sense-of-touch-to-paralysed-man.

[47] Mark Waghorn, "Robotic Arm Named After Luke Skywalker Enables Amputee to Touch and Feel Again," *Independent*, July 24, 2019, https://www.independent.co.uk/news/science/robotic-arm-luke-skywalker-amputee-prosthetic-university-utah-a9019211.html.

[48] Warrior Maven, "The Army Is Testing a New Super-Soldier Exoskeleton," *Business Insider*, November 27, 2017, https://www.businessinsider.com/army-testing-super-soldier-exoskeleton-2017-11.

[49] Richard Waters, "Exoskeletons to Become Common for Factory Workers," 50 Ideas to Change the World, video, 3:17, *Financial Times*, March 4, 2018, https://www.ft.com/content/4b5f7be2-1d6d-11e8-aaca-4574d7dabfb6.

50. Michael Winerip, "You Call That a Tomato?" *New York Times*, June 24, 2013, https://www.nytimes.com/2013/06/24/booming/you-call-that-a-tomato.html.

51. David Biello, "Genetically Modified Crop on the Loose and Evolving in U.S. Midwest," *Scientific American*, August 6, 2010, https://www.scientificamerican.com/article/genetically-modified-crop/.

52. Jennifer Ackerman, "Food: How Altered?" *National Geographic*, accessed June 24, 2020, https://www.nationalgeographic.com/environment/global-warming/food-how-altered/.

53. Associated Press, "Restaurants Could Be First to Get Genetically Modified Salmon," CBS News, June 21, 2019, https://www.cbsnews.com/news/restaurants-could-be-first-to-get-genetically-modified-salmon/.

54. Ed Yong, "New Zealand's War on Rats Could Change the World," *Atlantic*, November 2017, https://www.theatlantic.com/science/archive/2017/11/new-zealand-predator-free-2050-rats-gene-drive-ruh-roh/546011/.

55. Yong, "New Zealand's War on Rats Could Change the World."

56. Esvelt, Kevin, "'Gene Drives' Are Too Risky for Field Trials, Scientists Say," *New York Times*, November 16, 2017, https://www.nytimes.com/2017/11/16/science/gene-drives-crispr.html.

57. Knvul Sheikh, "Lab-Grown Meat That Doesn't Look Like Mush," *New York Times*, October 27, 2019, https://www.nytimes.com/2019/10/27/science/lab-meat-texture.html.

58. Zara Stone, "The High Cost of Lab-to-Table Meat," *Wired*, March 8, 2018, https://www.wired.com/story/the-high-cost-of-lab-to-table-meat/.

59. Jennifer Langston, "With a 'Hello' Microsoft and UW Demonstrate First Fully Automated DNA Data Storage," *Innovation Stories* (blog), Microsoft, March 21, 2019, https://news.microsoft.com/innovation-stories/hello-data-dna-storage/.

60. David Robson, "The Birth of Half-Human, Half-Animal Chimeras," BBC, January 5, 2017, http://www.bbc.com/earth/story/20170104-the-birth-of-the-human-animal-chimeras.

61. Nicola Davis, "Breakthrough as Scientists Grow Sheep Embryos Containing Human Cells," *Guardian*, February 17, 2018, https://www.theguardian.com/science/2018/feb/17/breakthrough-as-scientists-grow-sheep-embryos-containing-human-cells.

62. Robson, "The Birth of Half-Human, Half-Animal Chimeras."

63. Dyllan Furness, "Science of the Lambs: We Can Now Grow Human Cells in Sheep," *Digital Trends*, March 3, 2018, https://www.digitaltrends.com/cool-tech/scientists-created-human-sheep-chimera-embryos/.

64. Robin Seaton Jefferson, "Scientists 'Print' World's First Heart with Human Bioinks, Next 'Teach Them to Behave' like Hearts," *Forbes*, April 18, 2019, https://www.forbes.com/sites/robinseatonjefferson/2019/04/18/scientists-print-worlds-first-heart-with-human-bioinks-next-teach-them-to-behave-like-hearts/.

65. Robson, "The Birth of Half-Human, Half-Animal Chimeras."

66. Pallab Ghosh, "The GM Chickens That Lay Eggs with Anti-Cancer Drugs," BBC News, January 28, 2019, https://www.bbc.com/news/science-environment-46993649.

67 Robin McKie, "£1,000 for a Micro-pig. Chinese Lab Sells Genetically Modified Pets," *Guardian*, October 3, 2015, https://www.theguardian.com/world/2015/oct/03/micropig-animal-rights-genetics-china-pets-outrage.

68 Sigal Samuel, "A Celebrity Biohacker Who Sells DIY Gene-Editing Kits Is Under Investigation," Vox, May 19, 2019, https://www.vox.com/future-perfect/2019/5/19/18629771/biohacking-josiah-zayner-genetic-engineering-crispr.

69 Sarah Zhang, "A Biohacker Regrets Publicly Injecting Himself with CRISPR," *Atlantic*, February 20, 2018, https://www.theatlantic.com/science/archive/2018/02/biohacking-stunts-crispr/553511/.

70 Meg Tirrell, "A US Drugmaker Offers to Cure Rare Blindness for $850,000," CNBC, January 3, 2018, https://www.cnbc.com/2018/01/03/spark-therapeutics-luxturna-gene-therapy-will-cost-about-850000.html.

71 Rare Daily Staff, "Spark Prices Gene Therapy for Eye Disease at $850,000, Introduces Outcomes-Based Rebates," Global Genes, January 3, 2018, https://globalgenes.org/2018/01/03/spark-prices-gene-therapy-for-eye-disease-at-850000-introduces-outcomes-based-rebates/.

72 Jessica Lussenhop, "Why I Injected Myself with An Untested Gene Therapy," BBC News, November 21, 2017, https://www.bbc.com/news/world-us-canada-41990981.

73 Emily Mullin, "Before He Died, This Biohacker Was Planning a CRISPR Trial in Mexico," *MIT Technology Review*, May 4, 2018, https://www.technologyreview.com/2018/05/04/143034/before-he-died-this-biohacker-was-planning-a-crispr-trial-in-mexico/.

74 Zhang, "A Biohacker Regrets Publicly Injecting Himself with CRISPR."

75 Clyde Haberman, "Scientists Can Design 'Better' Babies. Should They?" *New York Times*, June 10, 2018, https://www.nytimes.com/2018/06/10/us/11retro-baby-genetics.html.

76 Ariana Eunjung Cha, "From Sex Selection to Surrogates, American IVF Clinics Provide Services Outlawed Elsewhere," *Washington Post*, December 30, 2018, https://www.washingtonpost.com/national/health-science/from-sex-selection-to-surrogates-american-ivf-clinics-provide-services-outlawed-elsewhere/2018/12/29/0b596668-03c0-11e9-9122-82e98f91ee6f_story.html.

77 Katie Moisse, "Fertility Clinic Offers Gender Selection, Draws Women from Abroad," ABC News, September 12, 2012, https://abcnews.go.com/Health/Wellness/fertility-clinic-offers-gender-selection-draws-women-abroad/story?id=17219176.

78 Cha, "From Sex Selection to Surrogates, American IVF Clinics Provide Services Outlawed Elsewhere."

79 Rob Stein, "Clinic Claims Success in Making Babies with 3 Parents' DNA," NPR, June 6, 2018, https://www.npr.org/sections/health-shots/2018/06/06/615909572/inside-the-ukrainian-clinic-making-3-parent-babies-for-women-who-are-infertile.

80 Rob Stein, "A Russian Biologist Wants to Create More Gene-Edited Babies," NPR, June 21, 2019, https://www.npr.org/sections/health-shots/2019/06/21/733782145/a-russian-biologist-wants-to-create-more-gene-edited-babies.

81 Stein, "A Russian Biologist Wants to Create More Gene-Edited Babies."

FORCE 3: HUMAN EXPANSIONISM

1. Bernard Marr, "15 Things Everyone Shoould Know About Quantum Computing," *Forbes*, October 10, 2017, https://www.forbes.com/sites/bernardmarr/2017/10/10/15-things-everyone-should-know-about-quantum-computing/?sh=51df69181f73.

2. Anne Trafton, "Team Invents Method to Shrink Objects to the Nanoscale," *MIT News*, December 13, 2018, http://news.mit.edu/2018/shrink-any-object-nanoscale-1213.

3. James Dennin, "'Clone Wars' Season 7 Trailer Shows a Famous Prequels Scene from a New Angle," *Inverse*, November 20, 2018, https://www.inverse.com/article/51056-new-nanochip-may-help-engineers-overcome-computing-limits-like-moore-s-law.

4. Luke Dormehl, "A Nanofiber Cloth Could Pull Fresh Drinking Water Straight from the Air," *Yahoo! News*, August 27, 2018, https://www.yahoo.com/news/nanofiber-cloth-could-pull-fresh-174532578.html.

5. Damon Cronshaw, "Nanotechnology That Promises to Save the World," *Newcastle Herald*, November 22, 2018, https://www.newcastleherald.com.au/story/5769894/dream-device-to-help-vehicles-run-on-sunlight-water-and-co2/.

6. Nanowerk News, "Nano-thin Invisibility Cloak Makes 3D Objects Disappear," *Nanowerk*, September 18, 2015, https://www.nanowerk.com/nanotechnology-news/newsid=41348.php.

7. Cell Press, "Nanotechnology Makes It Possible for Mice to See in Infrared," *Phys.org*, February 28, 2019, https://phys.org/news/2019-02-nanotechnology-mice-infrared.html.

8. Sam Million-Weaver, "It's Not a Shock: Better Bandage Promotes Powerful Healing," *University of Wisconsin-Madison News*, November 29, 2018, https://news.wisc.edu/its-not-a-shock-better-bandage-promotes-powerful-healing/.

9. Case Western Reserve University, "Case Western Reserve and Haima Therapeutics Sign Option License to Develop Synthoplate," *Daily*, June 4, 2018, https://thedaily.case.edu/case-western-reserve-haima-therapeutics-sign-option-license-develop-synthoplate/.

10. Dianne Price, "ASU Professor Named to Fast Company's 'Most Creative People in Business 2019,'" *ASU Alumni*, May 24, 2019, https://alumni.asu.edu/20190523-asu-news-hao-yan-named-fast-companys-most-creative-people-business-2019.

11. Joe Caspermeyer, "Cancer-Fighting Nanorobots Seek and Destroy Tumors," *ASU Now*, February 12, 2018, https://asunow.asu.edu/20180212-discoveries-cancer-fighting-nanorobots-seek-and-destroy-tumors.

12. Luke Dormehl, "Purdue's Microbots Are Designed to Wander Around Inside Your Body," *Digital Trends*, February 16, 2018, https://www.digitaltrends.com/cool-tech/microscale-tumbling-robots/.

13. Callum Hoare, "Asteroid Fears: NASA's Last-Ditch System in Place for Earth Impact ONE Week Away Exposed," *Express*, October 4, 2019, https://www.express.co.uk/news/science/1185456/asteroid-news-nasa-final-warning-earth-earth-impact-one-week-natalie-starkey-spt.

14. Leah Crane, "A Tech-Destroying Solar Flare Could Hit Earth Within 100 Years," *New Scientist*, October 16, 2017, https://www.newscientist.com/article/2150350-a-tech-destroying-solar-flare-could-hit-earth-within-100-years/.

15. Liz Gannes, "Tech Renaissance Man Elon Musk Talks Cars, Spaceships and Hyperloops at D11," *All Things Digital*, May 29, 2013, http://allthingsd.com/20130529/coming-up-tech-renaissance-man-elon-musk-at-d11/.

16. David Szondy, "Bad News for Mars-Bound Astronauts—Cosmic Rays Damage Your GI Tract," *New Atlas*, October 2, 2018, https://newatlas.com/mars-astronauts-cosmic-rays-intestines/56611/.

17. Irene Klotz, "Boiling Blood and Radiation: 5 Ways Mars Can Kill," Space.com, May 11, 2017, https://www.space.com/36800-five-ways-to-die-on-mars.html.

18. News4Jax.com Staff, "Senate Panel OKs Plan to Send Astronauts to Mars," *News4Jax*, September 21, 2016, https://www.news4jax.com/tech/2016/09/21/senate-panel-oks-plan-to-send-astronauts-to-mars/.

19. Karen Northon, "NASA Funds Demo of 3D-Printed Spacecraft Parts Made, Assembled in Orbit," *NASA*, July 12, 2019, last updated March 17, 2020, https://www.nasa.gov/press-release/nasa-funds-demo-of-3d-printed-spacecraft-parts-made-assembled-in-orbit.

20. "Our Mission," Blue Origin, accessed June 24, 2020, https://www.blueorigin.com/our-mission.

21. "ULA and Blue Origin Announce Partnership to Develop New American Rocket Engine," Blue Origin, September 17, 2014, https://www.blueorigin.com/news/ula-and-blue-origin-announce-partnership-to-develop-new-american-rocket-engine.

22. Jamie Carter, "Jeff Bezos: Reusable Rockets Will Let a Trillion People Colonise the Solar System," *TechRadar*, July 16, 2017, https://www.techradar.com/news/jeff-bezos-reusable-rockets-will-let-a-trillion-people-colonising-the-solar-system.

23. Ashley Strickland, "Astronauts on the Moon and Mars May Grow Their Homes There out of Mushrooms, Says NASA," CNN, January 17, 2020, https://www.cnn.com/2020/01/17/world/nasa-moon-mars-fungus-scn/index.html.

24. "ESA Opens Oxygen Plant—Making Air Out of Moondust," European Space Agency, January 17, 2020, http://www.esa.int/Enabling_Support/Space_Engineering_Technology/ESA_opens_oxygen_plant_making_air_out_of_moondust.

25. John Bowden, "'Building Blocks' for Life Discovered in 3-Billion-Year-Old Organic Matter on Mars," *Hill*, June 7, 2018, https://thehill.com/policy/technology/391228-building-blocks-for-life-discovered-in-3-billion-year-old-organic-matter-on.

26. Ian Sample, "Nasa Scientists Find Evidence of Flowing Water on Mars," *Guardian*, September 28, 2015, https://www.theguardian.com/science/2015/sep/28/nasa-scientists-find-evidence-flowing-water-mars.

27. Leah Crane, "Terraforming Mars with Strange Silica Blanket Could Let Plants Thrive," *New Scientist*, July 15, 2019, https://www.newscientist.com/article/2209746-terraforming-mars-with-strange-silica-blanket-could-let-plants-thrive/.

28. Charles Q. Choi, "How to Feed a Mars Colony of 1 Million People," Space.com, September 18, 2019, https://www.space.com/how-feed-one-million-mars-colonists.html.

29. Alexandra Lozovschi, "World's First Trillionaire Will Make Fortune in Outer Space, Claims Goldman Sachs," *The Inquisitr*, April 22, 2018, https://www.inquisitr.com/4874112/worlds-first-trillionaire-will-make-their-fortune-in-outer-space-claims-goldman-sachs/.

30. Atossa Araxia Abrahamian, "How the Asteroid-Mining Bubble Burst," *MIT Technology Review*, June 26, 2019, https://www.technologyreview.com/s/613758/asteroid-mining-bubble-burst-history/.

31. Abrahamian, "How the Asteroid-Mining Bubble Burst."

32. Abrahamian, "How the Asteroid-Mining Bubble Burst."

33. Abrahamian, "How the Asteroid-Mining Bubble Burst."

34. Abrahamian, "How the Asteroid-Mining Bubble Burst."

35. Daniel Oberhaus, "Astronomers Are Annoyed at a New Zealand Company That Launched a Disco Ball into Orbit," *Vice*, January 25, 2018, https://www.vice.com/en_us/article/kznvzw/rocket-lab-humanity-star-astronomers-space-junk.

36. Leah Crane, "SpaceX Starlink Satellites Could Be 'Existential Threat' to Astronomy," *New Scientist*, January 9, 2020, https://www.newscientist.com/article/2229643-spacex-starlink-satellites-could-be-existential-threat-to-astronomy/.

37. Neel V. Patel, "An Emotionally Intelligent AI Could Support Astronauts on a Trip to Mars," *MIT Technology Review*, January 14, 2020, https://www.technologyreview.com/2020/01/14/64990/an-emotionally-intelligent-ai-could-support-astronauts-on-a-trip-to-mars/.

38. Kenneth Chang, "Where's Our Warp Drive to the Stars?" *New York Times*, November 19, 2018, https://www.nytimes.com/2018/11/19/science/space-travel-physics.html.

39. Chang, "Where's Our Warp Drive to the Stars?"

40. University of Bristol, "First Chip-to-Chip Quantum Teleportation Harnessing Silicon Photonic Chip Fabrication," Phys.org, December 24, 2019, https://phys.org/news/2019-12-chip-to-chip-quantum-teleportation-harnessing-silicon.html.

41. University of Nottingham, "Research Sheds New Light on Intelligent Life Existing Across the Galaxy," EurekaAlert!, June 15, 2020, https://www.eurekalert.org/pub_releases/2020-06/uon-rsn061220.php.

42. Samantha Rolfe, "Could Unseen Aliens Exist Among Us?" RealClear Science, January 11, 2020, https://www.realclearscience.com/articles/2020/01/11/could_unseen_aliens_exist_among_us_111251.html.

43. Antonio Regalado, "Engineering the Perfect Astronaut," *MIT Technology Review*, April 14, 2017, https://www.technologyreview.com/2017/04/15/152545/engineering-the-perfect-astronaut/ (accessed on June 25, 2020).

44. Hannah Devlin, "Woolly Mammoth on Verge of Resurrection, Scientists Reveal," *Guardian*, February 16, 2017, https://www.theguardian.com/science/2017/feb/16/woolly-mammoth-resurrection-scientists.

45. Regalado, "Engineering the Perfect Astronaut."

46. Scott Solomon, "If Humans Gave Birth in Space, Babies Would Have Giant, Alien-Shaped Heads," *Business Insider*, July 23, 2019, https://www.businessinsider.com/humans-gave-birth-space-earth-giant-alien-heads-2019-7 (accessed on June 25, 2020).

47. Tom Ellis, "World's First Living Organism with Fully Redesigned DNA Created," *The Guardian*, May 15, 2019, https://www.theguardian.com/science/2019/may/15/cambridge

-scientists-create-worlds-first-living-organism-with-fully-redesigned-dna (accessed on June 25, 2020).

48. Madhan Tirumalai, "Bacterial Genetics Could Help Researchers Block Interplanetary Contamination," *The Scientist*, July 31, 2018, https://www.the-scientist.com/notebook/bacterial-genetics-could-help-researchers-block-interplanetary-contamination-64500 (accessed on June 25, 2020).

FORCE 4: DEEP AUTOMATION

1. Justin Scheck, Rory Jones, and Summer Said, "A Prince's $500 Billion Desert Dream: Flying Cars, Robot Dinosaurs and a Giant Artificial Moon," *Wall Street Journal*, July 25, 2019, https://www.wsj.com/articles/a-princes-500-billion-desert-dream-flying-cars-robot-dinosaurs-and-a-giant-artificial-moon-11564097568.

2. Nicole Kobie, "Malaysia's City of the Future Is an Uncanny Valley," *Wired*, March 22, 2016, https://www.wired.co.uk/article/forest-city-malaysia-report.

3. "Can AI Be a Fair Judge in Court? Estonia Thinks So," *Wired*, March 30, 2019, https://www.wired.com/story/can-ai-be-fair-judge-court-estonia-thinks-so/.

4. "Can AI Be a Fair Judge in Court? Estonia Thinks So," *Wired*.

5. Caroline Haskins, "Dozens of Cities Have Secretly Experimented with Predictive Policing Software," *Vice*, February 6, 2019, https://www.vice.com/en_us/article/d3m7jq/dozens-of-cities-have-secretly-experimented-with-predictive-policing-software.

6. Mark Smith, "Can We Predict When and Where a Crime Will Take Place?" BBC News, October 30, 2018, https://www.bbc.com/news/business-46017239.

7. James Vincent, "Security Robots Are Mobile Surveillance Devices, Not Human Replacements," The Verge, November 14, 2019, https://www.theverge.com/2019/11/14/20964584/knightscope-security-robot-guards-surveillance-devices-facial-recognition-numberplate-mobile-phone .

8. Ally Jarmanning, "Mass. State Police Tested Out Boston Dynamics' Spot the Robot Dog. Civil Liberties Advocates Want to Know More," WBUR News, November 25, 2019, https://www.wbur.org/news/2019/11/25/boston-dynamics-robot-dog-massachusetts-state-police.

9. Jarmanning, "Mass. State Police Tested Out Boston Dynamics' Spot the Robot Dog."

10. Anouk Vleugels, "How to Catch Criminals Through IoT and Predictive Software," TNW, November 1, 2018, https://thenextweb.com/the-next-police/2018/11/01/police-iot-ports-crime/.

11. Jeremy Chan, "All Eyes on the Future," *Asian Scientist Magazine*, February 13, 2019, https://www.asianscientist.com/2019/02/print/supercomputing-sensetime-lin-dahua/.

12. James Vincent, "Artificial Intelligence Is Going to Supercharge Surveillance," The Verge, January 23, 2018, https://www.theverge.com/2018/1/23/16907238/artificial-intelligence-surveillance-cameras-security.

13. Rachel England, "Chinese School Uses Facial Recognition to Make Kids Pay Attention," Engadget, May 17, 2018, https://www.engadget.com/2018/05/17/chinese-school-facial-recognition-kids-attention/.
14. Kaleigh Rogers, "What Constant Surveillance Does to Your Brain," *Vice*, November 14, 2018, https://www.vice.com/en_us/article/pa5d9g/what-constant-surveillance-does-to-your-brain.
15. Vincent, "Artificial Intelligence Is Going to Supercharge Surveillance."
16. Bill Siwicki, "Google AI Now Can Predict Cardiovascular Problems from Retinal Scans," MobiHealthNews, February 20, 2018, https://www.mobihealthnews.com/content/google-ai-now-can-predict-cardiovascular-problems-retinal-scans.
17. Benjamin Goggin, "Inside Facebook's Suicide Algorithm: Here's How the Company Uses Artificial Intelligence to Predict Your Mental State from Your Posts," *Business Insider*, January 6, 2019, https://www.businessinsider.com/facebook-is-using-ai-to-try-to-predict-if-youre-suicidal-2018-12.
18. Ahmed Elgammal, "Generating 'Art' by Learning About Styles and Deviating from Style Norms," *Medium*, June 25, 2017, https://medium.com/@ahmed_elgammal/generating-art-by-learning-about-styles-and-deviating-from-style-norms-8037a13ae027.
19. Brian Merchant, "The Poem That Passed the Turing Test," *Vice*, February 5, 2015, https://www.vice.com/en_us/article/vvbxxd/the-poem-that-passed-the-turing-test.
20. Peter Caranicas, "Artificial Intelligence Could One Day Determine Which Films Get Made," *Variety*, July 5, 2018, https://variety.com/2018/artisans/news/artificial-intelligence-hollywood-1202865540/.
21. Caranicas, "Artificial Intelligence Could One Day Determine Which Films Get Made."
22. Steve Rose, "'It's a War Between Technology and a Donkey'—How AI Is Shaking Up Hollywood," *Guardian*, January 16, 2020, https://www.theguardian.com/film/2020/jan/16/its-a-war-between-technology-and-a-donkey-how-ai-is-shaking-up-hollywood.
23. Lucy Jordan, "Inside the Lab That's Producing the First AI-Generated Pop Album," *Seeker*, April 13, 2017, https://www.seeker.com/tech/artificial-intelligence/inside-flow-machines-the-lab-thats-composing-the-first-ai-generated-pop-album.
24. Dani Deahl, "How AI-Generated Music Is Changing the Way Hits Are Made," The Verge, August 31, 2018, https://www.theverge.com/2018/8/31/17777008/artificial-intelligence-taryn-southern-amper-music.
25. Colm Gorey, "'I Realised Machine Learning Could Make My Musical Dreams Come True,'" Siliconrepublic.com, February 22, 2019, https://www.siliconrepublic.com/machines/maya-ackerman-alysia-ai-music.
26. Gorey, "'I Realised Machine Learning Could Make My Musical Dreams Come True.'"
27. Lucy Handley, "The 'World's First' A.I. News Anchor Has Gone Live in China," CNBC, November 9, 2018, https://www.cnbc.com/2018/11/09/the-worlds-first-ai-news-anchor-has-gone-live-in-china.html.
28. Rose, "'It's a War Between Technology and a Donkey.'"
29. Michael Ruiz, "Horror Game Bring to Light Will Utilize Heart Rate Monitor to 'Enhance' Your Experience," *DualShockers*, May 1, 2018, https://www.dualshockers.com/bring-to-light-red-meat-games/.

30. Asma Khalid, "A Dirty Word in the U.S., 'Automation' Is a Buzzword in China," WBUR, November 20, 2017, https://www.wbur.org/bostonomix/2017/11/20/china-automation.

31. IFR, "Robots: China Breaks Historic Records in Automation," IFR Press Releases, 2017, https://ifr.org/news/robots-china-breaks-historic-records-in-automation.

32. Henry Blodget, "CEO of Apple Partner Foxconn: 'Managing One Million Animals Gives Me a Headache,'" *Business Insider*, January 19, 2012, https://www.businessinsider.com/foxconn-animals-2012-1.

33. Sherisse Pham, "How Richard Liu Built JD.com into a $45 Billion Tech Giant," CNN, September 4, 2018, https://money.cnn.com/2018/09/04/technology/jd-com-ceo-richard-liu/index.html.

34. Saheli Roy Choudhury and Eunice Yoon, "JD.com Chief Richard Liu Sees Drone Delivery as the Way to Reach China's Rural Consumers," CNBC, June 18, 2017, https://www.cnbc.com/2017/06/18/jd-com-ceo-richard-liu-talks-drones-automation-and-logistics.html.

35. "Chinese Online Retailer JD.com's Plan to Diversify," *Wall Street Journal*, June 13, 2017, https://www.wsj.com/articles/chinese-online-retailer-jd-coms-plan-to-diversify-1497374520.

36. Dow Jones & Company, "Moving Up the Market," *WSJ D.Live Asia* (conference), June 14, 2017, https://images.dowjones.com/wp-content/uploads/sites/121/2017/12/04152646/DliveAsiaSpecialReport.pdf.

37. Peter Holley, "The Future of Autonomous Delivery May Be Unfolding in an Unlikely Place: Suburban Houston," *Washington Post*, November 13, 2019, https://www.washingtonpost.com/technology/2019/11/07/future-autonomous-delivery-may-be-unfolding-an-unlikely-place-suburban-houston/.

38. Bill Ibelle, "In The Future, Robots Will Perform Surgery, Shop for You, and Even Recycle Themselves," News@Northeastern, April 12, 2018, https://news.northeastern.edu/2018/04/12/in-the-future-robots-will-perform-surgery-shop-for-you-and-even-recycle-themselves/.

39. Ibelle, "In The Future, Robots Will Perform Surgery, Shop for You, and Even Recycle Themselves."

40. Elizabeth Svoboda, "Your Robot Surgeon Will See You Now," *Nature*, September 25, 2019, https://www.nature.com/articles/d41586-019-02874-0.

41. Michael Larkin, "Labor Terminators: Farming Robots Are About to Take Over Our Farms," *Investor's Business Daily*, August 10, 2018, https://www.investors.com/news/farming-robot-agriculture-technology/.

42. Gary Pullano, "Harvest CROO Robotics Strawberry Harvester Nears Fruition," *Fruit Growers News*, March 26, 2019, https://fruitgrowersnews.com/article/harvest-croo-robotics-strawberry-harvester-nears-fruition/.

43. John Seabrook, "The Age of Robot Farmers," *New Yorker*, April 15, 2019, https://www.newyorker.com/magazine/2019/04/15/the-age-of-robot-farmers.

44. Larkin, "Labor Terminators: Farming Robots Are About to Take Over Our Farms."

45. Lauren Comiteau, "World's 1st Floating Dairy Farm Could Help Cities Adapt to Climate Change," CBC News, December 13, 2019, https://www.cbc.ca/news/technology/floating-dairy-farm-1.5089424.

46. Simon Fry, "The World's First Floating Farm Making Waves in Rotterdam," BBC News, August 17, 2018, https://www.bbc.com/news/business-45130010.

47 Karen Graham, "Iron Ox Shows How AI and Robots Can Increase Farm Production," *Digital Journal*, October 3, 2019, http://www.digitaljournal.com/tech-and-science/technology/iron-ox-shows-us-how-ai-and-robots-can-increase-farm-production/article/533671.

48 Brandon Alexander, "If Farms Are to Survive, We Need to Think About Them as Tech Companies," *Quartz*, October 3, 2018, https://qz.com/1383635/if-farms-are-to-survive-we-need-to-think-about-them-as-tech-companies/.

49 Kazuaki Nagata, "Cyberdyne's HAL Suits Give Lift to Mobility-Challenged," *Japan Times*, July 13, 2014, https://www.japantimes.co.jp/news/2014/07/13/national/cyberdynes-hal-suits-give-lift-mobility-challenged/.

50 Jeevan Vasagar, "How Robots Are Teaching Singapore's Kids," *Financial Times*, July 12, 2017, https://www.ft.com/content/f3cbfada-668e-11e7-8526-7b38dcaef614.

51 Joseph Bennington-Castro, "Let Robots Teach Our Kids? Here's Why That Isn't Such a Bad Idea," NBC Mach, April 19, 2017, https://www.nbcnews.com/mach/technology/robots-will-soon-become-our-children-s-tutors-here-s-n748196.

52 Bryan Clark, "Rolls-Royce Is Working on a Robotic Cockroach That Can Fix Plane Engines," TNW, July 17, 2018, https://thenextwebcom/artificial-intelligence/2018/07/18/rolls-royce-is-working-on-robotic-cockroaches-that-fix-plane-engines/.

53 Brown University, "Dynamic Hydrogel Used to Make 'Soft Robot' Components and LEGO-like Building Blocks," News from Brown, March 21, 2019, https://www.brown.edu/news/2019-03-21/hydrogel.

54 Katyanna Quach, "Is That You, T-1000? No, Just a Lil Robot That Can Mimic Humans on Sight," *Register*, February 8, 2018, https://www.theregister.com/2018/02/08/robot_copycat_learning/.

FORCE 5: INTELLIGENCE EXPLOSION

1 Maureen Dowd, "Silicon Valley Sharknado," *New York Times*, July 8, 2014, https://www.nytimes.com/2014/07/09/opinion/maureen-dowd-silicon-valley-sharknado.html.

2 James Vincent, "This Is When AI's Top Researchers Think Artificial General Intelligence Will Be Achieved," The Verge, November 27, 2018, https://www.theverge.com/2018/11/27/18114362/ai-artificial-general-intelligence-when-achieved-martin-ford-book.

3 Catherine Clifford, "Elon Musk: 'Mark My Words—A.I. Is Far More Dangerous Than Nukes,'" CNBC, March 13, 2018, https://www.cnbc.com/2018/03/13/elon-musk-at-sxsw-a-i-is-more-dangerous-than-nuclear-weapons.html.

4 Vincent, "AI's Top Researchers."

5 Bec Crew, "This Physicist Says Consciousness Could Be a New State of Matter," Science Alert, September 16, 2016, https://www.sciencealert.com/this-physicist-is-arguing-that-consciousness-is-a-new-state-of-matter.

6 David Robson, "Giulio Tononi's 'Integrated Information Theory' Might Solve Neuroscience's Biggest Puzzle," BBC Future, March 26, 2019, https://www.bbc.com/future/article/20190326-are-we-close-to-solving-the-puzzle-of-consciousness.

7. Robert Lawrence Kuhn, "The Singularity, Virtual Immortality and the Trouble with Consciousness (Op-Ed)," Live Science, October 16, 2015, https://www.livescience.com/52503-is-it-possible-to-transfer-your-mind-into-a-computer.html.

8. Daniel Dennettl, "Is Superintelligence Impossible?" interview with John Brockman, Edge.org, April 10, 2019, https://www.edge.org/conversation/david_chalmers-daniel_c_dennett-on-possible-minds-philosophy-and-ai.

9. Jennifer Schussler, "Philosophy That Stirs the Waters," *New York Times*, April 29, 2013, https://www.nytimes.com/2013/04/30/books/daniel-dennett-author-of-intuition-pumps-and-other-tools-for-thinking.html.

10. Janosch Delcker, "Europe Divided over Robot 'Personhood,'" Politico, April 13, 2018, https://www.politico.eu/article/europe-divided-over-robot-ai-artificial-intelligence-personhood/.

11. Allison P. Davis, "Are We Ready for Robot Sex?" *New York Magazine*, May 14, 2018, https://www.thecut.com/2018/05/sex-robots-realbotix.html.

12. Breanna Mroczek, "Can Robots Feel Emotions?" *Disruption*, April 2019, https://www.disruptionmagazine.ca/can-robots-feel-emotions/.

13. Shane Schick, "Sanctuary AI Founder Is Making Robots That Are Exact Human Replicas (Starting with Herself)," B2B News Network, October 23, 2018, https://www.b2bnn.com/2018/10/sanctuary-ai-robotics/.

14. Hiroshi Ishiguro, "Sentient Love," interview, *52 Insights*, October 22, 2019, https://www.52-insights.com/hiroshi-ishiguro-sentient-love-robots-android-interview-technology/.

15. "MIT Helps Machine Learning Systems to Perceive Human Emotions," Internet of Business, https://internetofbusiness.com/mit-helps-machine-learning-systems-to-perceive-human-emotions/.

16. Adam Conner-Simons and Rachel Gordon, "Detecting Emotions with Wireless Signals," MIT News, September 20, 2016, http://news.mit.edu/2016/detecting-emotions-with-wireless-signals-0920.

17. Khari Johnson, "Softbank Robotics Enhances Pepper the Robot's Emotional Intelligence," VentureBeat, August 28, 2018, https://venturebeat.com/2018/08/28/softbank-robotics-enhances-pepper-the-robots-emotional-intelligence/.

18. Rachel Lerman, "Be Wary of Robot Emotions; 'Simulated Love Is Never Love,'" AP News, April 26, 2019, https://apnews.com/99c9ec8ebad242ca88178e22c7642648.

19. Joelle Renstrom, "Will Care-Bots Cure the Loneliness of Nursing Homes?" *Daily Beast*, May 1, 2019, https://www.thedailybeast.com/will-care-bots-cure-the-loneliness-of-nursing-homes.

20. David Cox, "Can These Little Robots Ease the Big Eldercare Crunch?" NBC Mach, November 12, 2017, https://www.nbcnews.com/mach/science/can-these-little-robots-ease-big-eldercare-crunch-ncna819841.

21. Cox, "Can These Little Robots Ease the Big Eldercare Crunch?"

22. Malcolm Foster, "Aging Japan: Robots May Have Role in Future of Elder Care," Reuters, March 27, 2018, https://www.reuters.com/article/us-japan-ageing-robots-widerimage/aging-japan-robots-may-have-role-in-future-of-elder-care-idUSKBN1H33AB.

23. Foster, "Aging Japan: Robots May Have Role in Future of Elder Care."

24. "Scientists Develop Robot Personal Trainer to Coach at Gym," *Irish News*, November 29, 2019, https://www.irishnews.com/magazine/daily/2019/11/29/news/scientists-develop-robot-personal-trainer-to-coach-at-gym-1778566/.

25. "Scientists Develop Robot Personal Trainer to Coach at Gym."

26. James Vincent, "New Study Finds It's Harder to Turn Off a Robot When It's Begging for Its Life," The Verge, August 2, 2018, https://www.theverge.com/2018/8/2/17642868/robots-turn-off-beg-not-to-empathy-media-equation.

27. Vincent, "New Study Finds It's Harder to Turn Off a Robot When It's Begging for Its Life."

28. Evan Ackerman, "Do Kids Care If Their Robot Friend Gets Stuffed into a Closet?" IEEE Spectrum, April 30, 2012, https://spectrum.ieee.org/automaton/robotics/artificial-intelligence/do-kids-care-if-their-robot-friend-gets-stuffed-into-a-closet.

29. Ackerman, "Do Kids Care If Their Robot Friend Gets Stuffed into a Closet?"

30. John Loeffler, "Robot on the Trolley Car Track: How Valuable Is Robot Life?" *Interesting Engineering*, February 9, 2019, https://interestingengineering.com/robot-on-the-trolley-car-track-how-valuable-is-robot-life.

31. Lerman, "Be Wary of Robot Emotions; 'Simulated Love Is Never Love.'"

32. Tom Hoggins, "Could You Fall in Love with a Robot? It May Be More Likely Than You Think," *Telegraph*, May 7, 2019, https://www.telegraph.co.uk/technology/2019/05/07/could-fall-love-robot-romance-machines-could-future-relationships/.

33. Alex Williams, "Do You Take This Robot . . . ," *New York Times*, January 19, 2019, https://www.nytimes.com/2019/01/19/style/sex-robots.html.

34. Davis, "Are We Ready for Robot Sex?"

35. Williams, "Do You Take This Robot . . ."

36. Williams, "Do You Take This Robot . . ."

37. Jack Schofield, "Let's Talk About Sex . . . with Robots," *Guardian*, September 16, 2009, https://www.theguardian.com/technology/2009/sep/16/sex-robots-david-levy-loebner.

38. "Interview Sherry Turkle," *FRONTLINE*, September 22, 2009, https://www.pbs.org/wgbh/pages/frontline/digitalnation/interviews/turkle.html.

39. Brandon Ambrosino, "Are We Set for a New Sexual Revolution?" BBC Future, July 2, 2019, https://www.bbc.com/future/article/20190702-are-we-set-for-a-new-sexual-revolution.

40. Eva Wiseman, "Sex, Love and Robots: Is This the End of Intimacy?" *Guardian*, December 13, 2015, https://www.theguardian.com/technology/2015/dec/13/sex-love-and-robots-the-end-of-intimacy.

41. Andrea Morris, "Meet the Activist Fighting Sex Robots," *Forbes*, September 26, 2018, https://www.forbes.com/sites/andreamorris/2018/09/26/meet-the-activist-fighting-sex-robots/.

42. Wynne Parry, "Robot Manipulates Humans in Creepy New Experiment. Should We Be Worried?" NBC Mach, August 14, 2018, https://www.nbcnews.com/mach/science/robot-manipulates-humans-creepy-new-experiment-should-we-be-worried-ncna900361.

43. Tahira Noor Khan, "69% of Managers' Work to Be Completely Automated by 2024: Gartner," *Entrepreneur*, January 23, 2020, https://www.entrepreneur.com/article/345435.

44. "New Study: 64% of People Trust a Robot More Than Their Manager," Oracle, October 15, 2019, https://www.oracle.com/corporate/pressrelease/robots-at-work-101519.html.

45. James Vincent, "DeepMind's Go-Playing AI Doesn't Need Human Help to Beat Us Anymore," The Verge, October 18, 2017, https://www.theverge.com/2017/10/18/16495548/deepmind-ai-go-alphago-zero-self-taught.

46. Tristan Greene, "Hyperdimensional Computing Theory Could Lead to AI with Memories and Reflexes," TNW, May 17, 2019, https://thenextweb.com/artificial-intelligence/2019/05/17/hyperdimensional-computing-theory-could-lead-to-ai-with-memories-and-reflexes/.

47. Alyssa Foote, "Robot 'Natural Selection' Recombines into Something Totally New," *Wired*, March 26, 2019, https://www.wired.com/story/how-we-reproduce-robots/.

48. Foote, "Robot 'Natural Selection' Recombines into Something Totally New."

49. Tristan Greene, "One Machine to Rule Them All: A 'Master Algorithm' May Emerge Sooner Than You Think," TNW, April 17, 2018, https://thenextweb.com/artificial-intelligence/2018/04/17/one-machine-to-rule-them-all-a-master-algorithm-may-emerge-sooner-than-you-think/.

50. Stephen Nellis, "Microsoft to Invest $1 Billion in OpenAI," Reuters, July 22, 2019, https://www.reuters.com/article/us-microsoft-openai/microsoft-to-invest-1-billion-in-openai-idUSKCN1UH1H9.

51. "Can Technology Plan Economies and Destroy Democracy?" *Economist*, December 18, 2019, https://www.economist.com/christmas-specials/2019/12/18/can-technology-plan-economies-and-destroy-democracy.

52. Ariel Bleicher, "Demystifying the Black Box That Is AI," *Scientific American*, August 9, 2017, https://www.scientificamerican.com/article/demystifying-the-black-box-that-is-ai/.

53. Vincent Elkaim, "Google's AI Guru Wants Computers to Think More Like Brains," *Wired*, December 12, 2018, https://www.wired.com/story/googles-ai-guru-computers-think-more-like-brains/.

54. Clifford, "Elon Musk: 'Mark My Words—A.I. Is Far More Dangerous Than Nukes.'"

55. Rory Cellan-Jones, "Stephen Hawking Warns Artificial Intelligence Could End Mankind," BBC News, December 2, 2014, https://www.bbc.com/news/technology-30290540.

56. Maureen Dowd, "Elon Musk's Billion-Dollar Crusade to Stop the A.I. Apocalypse," *Vanity Fair*, April 2017, https://www.vanityfair.com/news/2017/03/elon-musk-billion-dollar-crusade-to-stop-ai-space-x.

57. Emilio Calvano, Giacomo Calzolari, Vincenzo Denicolò, and Sergio Pastorello, "Artificial Intelligence, Algorithmic Pricing, and Collusion," VoxEU.org, February 3, 2019, https://voxeu.org/article/artificial-intelligence-algorithmic-pricing-and-collusion.

58. Dan Robitzski, "Five Experts Share What Scares Them the Most About AI," *Futurism*, September 5, 2016, https://futurism.com/artificial-intelligence-experts-fear.

59. Henry McDonald, "Ex-Google Worker Fears 'Killer Robots' Could Cause Mass Atrocities," *Guardian*, September 15, 2019, https://www.theguardian.com/technology/2019/sep/15/ex-google-worker-fears-killer-robots-cause-mass-atrocities.

60. Cristianna Reedy, "Kurzweil Claims That the Singularity Will Happen by 2045," *Futurism*, October 5, 2017, https://futurism.com/kurzweil-claims-that-the-singularity-will-happen-by-2045.

61. Cadie Thompson, "Live Forever? Maybe, by Uploading Your Brain," CNBC, May 4, 2015, https://www.cnbc.com/2015/05/04/live-forever-maybe-by-uploading-your-brain.html.

62. Adam Piore, "The Neuroscientist Who Wants to Upload Humanity to a Computer," *Popular Science*, May 16, 2014, https://www.popsci.com/article/science/neuroscientist-who-wants-upload-humanity-computer/.

63. Piore, "The Neuroscientist Who Wants to Upload Humanity to a Computer."

64. Scott Fulton III, "Neuromorphic Computing and the Brain That Wouldn't Die," ZDNet, February 27, 2019, https://www.zdnet.com/article/neuromorphic-computing-and-the-brain-that-wouldnt-die/.

65. Luke Dormehl, "Start-up Can Preserve Your Brain for Future Upload, but It's '100 Percent Fatal,'" Digital Trends, March 13, 2018, https://www.digitaltrends.com/cool-tech/nectome-brain-embalm-mind-uploading/.

66. Antonio Regalado, "A Start-up Is Pitching a Mind-Uploading Service That Is '100 Percent Fatal,'" *MIT Technology Review*, March 13, 2018, https://www.technologyreview.com/2018/03/13/144721/a-startup-is-pitching-a-mind-uploading-service-that-is-100-percent-fatal/#.

67. Yohan John, "What Percent Chance Is There That Whole Brain Emulation or Mind Uploading to a Neural Prosthetic Will Be Feasible by 2048?" Quora, December 11, 2013, https://www.quora.com/What-percent-chance-is-there-that-whole-brain-emulation-or-mind-uploading-to-a-neural-prosthetic-will-be-feasible-by-2048.

68. Beth Elderkin, "Will We Ever Be Able to Upload a Mind to a New Body?" Gizmodo, February 5, 2018, https://gizmodo.com/will-we-ever-be-able-to-upload-a-mind-to-a-new-body-1822622161/amp.

69. Johnthomas Didymus, "Google's Ray Kurzweil: 'Mind Upload' Digital Immortality by 2045," *Digital Journal*, June 20, 2013, http://www.digitaljournal.com/article/352787.

70. Hannah Devlin, "Monkey 'Brain Net' Raises Prospect of Human Brain-to-Brain Connection," *Guardian*, July 9, 2015, https://www.theguardian.com/science/2015/jul/09/monkey-brain-net-raises-prospect-of-human-brain-to-brain-connection.

71. Jason Koebler, "A Researcher Made an Organic Computer Using Four Wired-Together Rat Brains," *Vice*, July 9, 2015, https://www.vice.com/en_us/article/bmj49v/a-researcher-made-an-organic-computer-using-four-wired-together-rat-brains.

72. Randy Rieland, "Scientists Connect Monkey Brains and Boost Their Thinking Power," *Smithsonian Magazine*, July 20, 2015, https://www.smithsonianmag.com/innovation/scientists-connect-monkey-brains-and-boost-their-thinking-power-180955963/.

73. Koebler, "A Researcher Made an Organic Computer Using Four Wired-Together Rat Brains."

74. Will Knight, "Enhanced Intelligence, VR Sex, and Our Cyborg Future," *Wired*, December 30, 2019, https://www.wired.com/story/enhanced-intelligence-vr-sex-our-cyborg-future/.

NOTES

75 Michelle Z. Donahue, "How a Color-Blind Artist Became the World's First Cyborg," *National Geographic*, April 3, 2017, https://www.nationalgeographic.com/news/2017/04/worlds-first-cyborg-human-evolution-science/.

76 Donahue, "How a Color-Blind Artist Became the World's First Cyborg."

77 Michael Abrams and Dan Winters, "Can You See with Your Tongue?" *Discover*, May 31, 2003, https://www.discovermagazine.com/mind/can-you-see-with-your-tongue.

78 Mandy Kendrick, "Tasting the Light: Device Lets the Blind 'See' with Their Tongues," *Scientific American*, August 13, 2009, https://www.scientificamerican.com/article/device-lets-blind-see-with-tongues/.

79 Kendrick, "Tasting the Light."

80 Erik Weihenmayer, "Seeing with Your Tongue," *New Yorker*, May 15, 2017, https://www.newyorker.com/magazine/2017/05/15/seeing-with-your-tongue.

81 Claire Maldarelli, "A Device for the Deaf That Lets You 'Listen' with Your Skin," *Popular Science*, September 30, 2016, https://www.popsci.com/device-that-lets-you-listen-with-your-skin/.

82 Burkhard Bilger, "The Possibilian," *New Yorker*, April 25, 2011, https://www.newyorker.com/magazine/2011/04/25/the-possibilian.

83 Bilger, "The Possibilian."

84 David Grossman, "Scientists Re-create Baby Brain Readings in a Dish," *Popular Mechanics*, Nov 19, 2019, https://www.popularmechanics.com/science/animals/a25224015/lab-brain-tissue-human-brain-waves/.

85 "Robot with a Biological Brain: New Research Provides Insights into How the Brain Works," University of Reading, August 14, 2008, https://www.reading.ac.uk/news-archive/press-releases/pr16530.html.

INDEX

A

Ackerman, Maya, 168
actors, virtual, 171
Adams, Scott, 28–29
Adleman, Leonard, 78
adversarial network, 159
advertising, BCIs and, 6
AeroFarms, 184
Affectiva, 207–208, 210
AGI (artificial general intelligence). *See* superintelligence
aging, 51–59
agriculture, 180–185. *See also* food; livestock
AI (artificial intelligence). *See* artificial intelligence (AI)
AIVA, 168
Alcaide, Ramses, 5–6
Alcor Life Extension Foundation, 61
Alcubierre, Miguel, 126
ALE, 122
Alexa, 18, 150, 217
Alexander, Brandon, 184–185
Alibaba, 189, 226
aliens, 128–129
AlphaGo, 220–221
AlphaGo Zero, 221
Alta Charo, R., 92
AlterEgo, 12–13, 18
Altman, Sam, 224, 238
ALYSIA, 168
Amazon, 175
    data collected by, 165, 226
    prediction by, 150
Ambrosia, 54–55
American Civil Liberties Union, 142, 148
American Medical Association (AMA), 46
Amper, 168
Anderson, Chad, 121
ANI (artificial narrow intelligence), 134, 196, 228–229
animals. *See also* livestock
    cloning, 62–64, 65
    genetically altered pets, 85
    new, 40, 72, 129–130
anonymity, 145. *See also* privacy
Ant-Man, 102
Appel, Larry, 50
AquaBounty, 71
AR (augmented reality). *See* augmented reality (AR)
Arnoldussen, Aimee, 250
ARPANET, 2
art/design, 39, 159–164
artificial general intelligence (AGI), 134. *See also* superintelligence
artificial intelligence (AI), xx
    automation and, 133. *see also* deep automation
    biases in, 141, 144, 166–167
    brainwaves and, 5. *see also* brain-computer interfaces (BCIs)
    dangers of, 229–235
    emotional, 124–125
    exoskeletons and, 68
    quantum computing and, 97
    responding to, 233
    safeguards for, 230–235
artificial narrow intelligence (narrow AI/ANI), 134, 196, 228–229

INDEX

Ascendance Biomedical, 88
Asimov, Isaac, 231–232
Asprey, Dave, 48–49
asteroids, 112, 120–122
augmented reality (AR), 3, 30–33, 35, 155
automation, xx, 134, 192–194. *See also* deep automation; robots
autonomy, 152–153, 154
Azermai, Nadira, 166, 172

B

Babitz, Liviu, 43
Bach-y-Rita, Paul, 249
bacterial spores, 131
Barcelona, 137, 138
Barzilai, Nir, 56, 57
BCIs (brain-computer interfaces). *See* brain-computer interfaces (BCIs)
Bedford, Robert, 60
Benioff, Paul, 96
Berger, Hans, 4
Berger, Theodore, 14, 236
Beridze, Irakli, 233
Bezos, Jeff, 117
Bhaskaran, Madhu, 104
biases in AI, 141, 144, 166–167
Bier, August, 41
Bigelow, Robert, 117–118
bin Salman, Mohammed, 135
Bing Su, 83
bio convergence, xx. *See also* biohacking; gene editing/genetic engineering
    bionics, 66–68
    chimeras, 80–85
    cloning, 62–66
    lab-grown meats, 76
    life extension, 51–59
    mind hacking, 37
    organic computers, 78–79
    potential of, 260
    smart drugs, 47–51
BioCurious, 39
Bioeconomy Capital, 39
biohacking, 37. *See also* bio convergence; transhumanism
    activists, 88–89
    antiaging, 51–59
    art/design and, 39–40
    body augmentation, 41–42
    body modification, 44–47
    cost of, 39–40
    CRISPR kits, 86
    cryonics, 60–62
    described, 38–40
    future of, 47
    legality of, 46
    motivations for, 40–41, 44
    nootropics, 47–51
    sensory augmentation, 43
    transhumanism, 38
biological computing devices, 78–79
biology, merging with technology, xx. *See also* bio convergence
bionics, 66–68
bioprinting, 82–83
Black Mirror, 147
Blade Runner, 167
blindness. *See* vision
blockchain, 225
blood plasma, 55
Blue Origin, 116, 117, 124, 125
body modification, 44–47. *See also* biohacking
Bongard, Joshua, 221–222
bosses, robot, 218–220
Bostrom, Nick, 29
Botstein, David, 52
Boulud, Daniel, 190
Boyalife Group, 64
Boyden, Edward, 238
Bradford Space, 121
brain chips, 8–15, 26–27, 35. *See also* brain-computer interfaces (BCIs)
Brain Music, 170
brain nets, 239–241, 245, 252–253
brain transplant, 65–66
brain-computer interfaces (BCIs), 3, 9–10, 35, 163, 241
    applications of, 15–16
    augmenting senses, 247–253
    brain nets, 239–241, 245, 252–253
    connected to robot bodies, 252
    goals of, 20
    privacy and, 17–18
    subconscious and, 241

INDEX

technologies being tested, 4–15
   use of data from, 17–18, 19
BrainGate, 10–11
brain-machine interfaces, 9–15
BrainPort, 249–250
brains
   artificial hippocampus, 14
   connecting to internet, 242–247
   conscious thought process, 240–241
   freezing, 237–238
   function of, 9
   growing, 253–256
   linking AI to, 11
   plasticity of, 8, 249
   preserving, 60–62
   reality and, 247
   subconscious, 241, 242
   synthetic neocortex, 27
   transferring thoughts, 10
brain-simulation technology, 236–237
brainwaves, 4–8, 13, 202
Braude, Peter, 65
Breithaupt, Fritz, 217
Brin, Sergey, 156
Brown, Louise, 65
Buehrer, Daniel J., 224
bugs, as food, 119–120
Bulletproof Coffee, 48

C

caloric restriction, 57
Calos, Michele, 88
Cambiaso, Adolfo, 63
Cambrian Genomics, 40
cameras, CCTV, 145, 146
Cameron, James, 121
Canavero, Sergio, 65–66
cancer, 79, 107–108. *See also* health care
Cannon, Tim, 41, 45
Cappelleri, David, 109
Carlson, Jeff, 109
Carlson, Rob, 39
cars
   AR navigation, 31
   flying, 137
   self-driving, 227
CCTV cameras, 145, 146

Chang, Edward, 11
Chase, Henry, 157
chimeras, 80–85
China
   AI news anchor in, 172
   digital currency in, 225
   manufacturing in, 174–175
   online education in, 155
   smart cities in, 135–136
   surveillance in, 147–148
Chisholm, Brock, 148
choices, predicting, 150–154
Church, George, 53–54, 129–130
Cinelytic, 165
cities, smart, 135–138
civil rights, 144
Clark, Gregory, 67
Clark, Luke, 157
Clark, Steve, 140
climate change. *See* environment/climate change
cloning, 62–66
Code Academy, 154
coding, 198–199
CognitionX, 167
Cole, Bryony, 214
college admission, 158
communications revolutions, 1–2. *See also* internet; mass connectivity
COMPAS, 140
computer networks, 134
computers, organic, 78–79
Connecterra, 182
conscious thought process, 240–241
consciousness, 9, 199–202. *See also* self-awareness
consciousness, collective, 245
Conselice, Christopher, 128
ConsenSys, 121
construction robots, 187–188
Copeland, Nathan, 67
corporate personhood, 204
Cortica, 146
Cosset, Yael, 176
COVID-19 pandemic, online learning and, 155
creativity
   art, 159–164

INDEX

gene editing and, 93
crime. *See* judges; police; security; surveillance
Crisanti, Andrea, 75
CRISPR, 69–76, 79, 82, 86, 87, 88, 89
Crockford, Kade, 142
CROO Robotics, 181
cryonics, 60–62
cryptocurrency, 225–226
Cuban Missile Crisis, 111
curiosity, 110
currency, digital, 225–226
Currin, Dawn, 47
Cyberdyne, 185–186
cyborgs, 42, 251

D

Darnovsky, Marcy, 91
DARPA (Defense Advanced Research Projects Agency), 14, 75
data. *See also* privacy
  abuse of, 17
  access to, 165, 226
  DNA storage of, 79
  use of, 19
Datta, Kamal, 114
de Grey, Aubrey, 61
De Maria, Lawrence, 182
Dean, James, 171
debate, 154
decisions, predicting, 150–154
deep automation, xx. *See also* automation; robotics; robots
  agriculture and, 180–185
  art and, 159–164
  education and, 154–159
  entertainment and, 164–174
  health care and, 177–180
  job losses and, 170–171
  jobless society and, 192–194
  manufacturing and, 174–175
  police and, 140–143
  potential of, 259
  prediction by, 149–154
  smart cities, 135–138
  smart governments, 138–140
  social order and, 134
  surveillance and, 143–149

Deep Space Industries, 121
deep-learning algorithms, 6
DeepStory, 172
Defense, US Department of, 14–15, 75. *See also* military
Defense Advanced Research Projects Agency (DARPA), 14, 75
Degray, Dennis, 11
delivery, retail, 175–176
Dennett, Daniel, 201, 202
dentists, robotic, 190
Descartes Labs, 180
determinism, 202
Dick, Philip K., 140
dictators, 205, 229
digisexual, 214
disease/illness, 54. *See also* drugs; health care; mental health
  biocircuits and, 79
  cancer, 79, 107–108
  chimeras and, 83
  cloning and, 62–63
  DIY treatments, 88
  drug-resistant, 72
  eliminating, 58
  organoids and, 253, 255
Disney, Walt, 60
DIY bio, 39
DNA, redesigned, 131
DNA computing, 78
dopamine, 157
Dorsey, Jack, 57
dreams, recording, 8
Driesch, Hans, 62
drones, 141, 176
drugs, 47–51, 84, 87–88, 103. *See also* health care
Duchamp, Marcel, 160
Duolingo, 154
Dvir, Tal, 82

E

Eagleman, David, 250–251
Earth, 110, 111–113. *See also* environment/climate change
economic singularity, 191–194
economies, AI, 225–228
economy, planned, 226

INDEX

Edmond de Belamy, 160, 161
education
    BCIs in, 7
    deep automation and, 154–159
    internet-connected brains and, 243
    online, 155
    robotic aides, 187
    special needs, 187
    surveillance in, 147–148
    virtual reality in, 155
EEG (electroencephalogram), 4–8, 13
Egenesis, 82
Egli, Dieter, 92
Einstein, Albert, 127
ejaculation, abstinence from, 49
El Kaliouby, Rana, 208
eldercare, 185, 209–210
electroencephalogram (EEG), 4–8, 13
Elgammal, Ahmed, 160
Ellis, Tom, 131
Ellison, Larry, 52
Elysium, 57
Emondi, Al, 14
emotional intelligence, 124–125, 196, 207–211
emotions, 243–244. *See also* love
empathy, 196, 208, 216
encryption, 98
End of Sex and The Future of Human Reproduction, The, 216
energy, 101, 137
Engstrom, David, 139
entertainment, deep automation and, 164–174
environment/climate change, 112
    drinkable water, 104–105
    GMO crops and, 70
    greenhouse gases, 105
    lab-based meats and, 77
e-pavement, 136–137
epidural electronics, 13
Esterhuizen, Jason, 25
Estonia, 138–140
Esvelt, Kevin, 74–75
e-tattoos, 13
eugenics, 90
European Space Agency, 118
European Union, 204
    Human Brain Project, 236

evolution, 53
Ex Machina, 218
exoskeletons, 68
extinction, human, 111–113
extinction events, 110

F

face transplants, 66
Facebook
    AI Research Lab, 159
    brain-computer interface, 18
    Building 8, 13
    data and, 17, 226
    prediction by, 151–152
facial expressions, 207
Fahy, Greg, 55
Fantastic Voyage, 109
farming, 180–185
FarmLogs, 180
FDA, 89
Feynman, Richard, 96
Ffirth, Stubbins, 41
film industry, 165–174
Finn, Kerry, 66–67
fitness, 210
five forces, xx. *See also* bio convergence; deep automation; human expansionism; intelligence explosion; mass connectivity
    potential of, 259–261
Floating Farm, 183–184
FlyZoo, 189
fMRI (functional magnetic resonance imaging), 7–8, 202
focus groups, 6
Foo Hui Hui, 187
food. *See also* livestock
    agriculture, 180–185
    bugs as, 119–120
    cloned meat, 62–63, 64
    genetically engineered, 69, 184
    lab-grown meats, 76–78
foot binding, 45–46
Ford, Martin, 197
Forest City, 136
FORTIS, 68
Fountain, 160, 161
Foxconn, 174–175

free will, 152–153, 154
Freezing People Is (Not) Easy, 60
functional magnetic resonance imaging (fMRI), 7–8, 202
functionalism, 201
Future of Sex, 214
Future Workplace, 219

G

games, 157–158, 173
Garg, Animesh, 179
Gartner, 219
Gates, Bill, 156
Bill and Melinda Gates Foundation, 103
Geisinger, 151
Gemma, Joe, 174
gene drives, 73–76
gene editing/genetic engineering, xx, 37, 38, 86. *See also* bio convergence; genetically modified organisms (GMOs)
   access to, 91–93
   for alien environments, 129
   banning, 73
   chimeras, 80–85
   CRISPR, 69–76, 79, 82, 86, 87, 88, 89
   food, 69
   gene drives, 73–76
   genetically modified animals, 129–130
   kits for, 39
   lab-grown meats, 76–78
   potential of, 260
   TALENS, 85
   of unborn children, 89–92
gene therapy, 87–88
genetic diversity, 62–63, 65
genetic engineering. *See* bio convergence; gene editing/genetic engineering
genetically modified organisms (GMOs), 69, 184. *See also* gene editing/genetic engineering
genomics, 69
Gerrol, Spencer, 6
Gildert, Suzanne, 205–207
Gimzewski, James, 236, 237
Global Slavery Index, 229
glove, haptic, 23
GMOs (genetically modified organisms), 69, 184

Go, 220–221
Goldman Sachs, 120
Goldstaub, Tabitha, 167
Google
   AI diagnostics, 177
   AR research, 32
   Calico, 51
   DeepMind, 150–151, 220–221
   Magenta, 168
   military projects, 233, 234
   quantum computing, 97
Google Glass, 30
Google Home, 217
Google Robotics, 221
Gorlitsky, Adam, 68
Gou, Terry, 174–175
governments
   AI and, 205
   BCIs and, 19
   smart, 138–140
graphene, 100–101
Gratch, Jonathan, 213
greenhouse gases, 105. *See also* environment/climate change
Grindhouse Wetware, 41, 45
Guarente, Leonard, 57
gyms, robots at, 210

H

hand transplants, 66
Handmaid's Tale, The, 64
Hanson, Jaydee, 63
Harbisson, Neil, 248–249
Hawking, Stephen, 230
He Jiankui, 89
health, surveillance and, 148
health care. *See also* disease/illness; eldercare; medicines
   AMA, 46
   AR in, 31–32
   BCIs in, 7
   deep automation and, 177–180, 190–191
   mental health, 148, 179–180, 207, 215–216
   nanotechnology and, 106–109
   predicting problems, 150–151
   quadriplegics, 10–11, 67
health span, 51–59

INDEX

hearing, 24, 250–251
heaven, 260
Heilig, Morton, 22, 23
Heinz, Austen, 40
Hellboy, 165
Hendricks, Michael, 61, 238
herbs, 50–51
Herron, Lissa, 84
Hertzfeld, Henry, 121
hibernation chambers, 125
Hill, Napoleon, 49
Hinton, Geoffrey, 227
hippocampus, artificial, 14
hiring, 158, 218–220
hive mind, 242–247
homeschooling, 158
Horvath, Steve, 55
Howard, David, 223
HR departments, 219
human expansionism, xx, 95. *See also*
    Mars; quantum computing; space
    asteroid mining, 120–122
    debate over, 132
    interplanetary ecosystem, 116–120
    nanotechnology, 102–109
    new materials, 99–102
    space economy, 123–125
human genomes, synthetic, 130
human trials, 56, 57, 58, 84, 89
Humanity Star, 122
humanness, 237. *See also* consciousness;
    identity
Hwang Woo-suk, 64
Hyman, Steve, 66
hyperdimensional computing
    theory, 222
HyperSciences, 122
Hyun, Insoo, 82

I

I, Robot, 231–232
IBM
    Deep Blue AI, 97
    Project Debater, 154
    quantum computing, 97
    Watson Beat, 168
ice ages, 110
identity, 236
immortality, 61

in vitro fertilization (IVF), 65
income, universal basic, 192
income inequality, 192
industry. *See also* manufacturing
    AR in, 31
    exoskeletons in, 68
inequality, income, 192
innovation, 2, 93
intelligence explosion, xx. *See also*
    superintelligence
internet, 2. *See also* mass connectivity
    connecting brain to, 242–247
internet, quantum, 128
internet of things (IoT), 134, 182
internet service, 122–123
intimacy, 42. *See also* sex
    with robots, 214, 215, 217
invasive species, 73–74
IoT (internet of things), 134, 182
IQ, 48, 91
Iron Ox, 184–185
Ishiguro, Hiroshi, 206, 248
Island of Doctor Moreau, The, 80
Izpisua Belmonte, Juan Carlos, 81, 83, 84

J

James, Watson, 56
Japan
    construction robots in, 187
    eldercare robots in, 210
    need for robots, 185
jet engines, 188
Jiajia, Zheng, 213
Jingdong, 175–176
job losses, deep automation and, 170–171
jobless society, possibility of, 192–194
Journal of Medical Ethics, 61
judges
    AI, 139
    automation and, 140–141
Jukebox, 169

K

Kac, Eduardo, 40
Kaeberlein, Matt, 56
Kahneman, Daniel, 241
Kapur, Arnav, 12–13
Karmazin, Jesse, 54–55

INDEX

Kasparov, Garry, 97
Katabi, Dina, 207
kava kava, 50–51
Kazerooni, Homayoon, 68
Kell, James, 188
Kessler, Donald, 123
Khan Academy, 154
Khosla, Vinod, 179
Knightscope, 141
Koch, Christof, 201
Koene, Randal, 236
Komar, Scott, 181
Kondo, Akihiko, 213
Konrath, Sara, 216
Körber Supply Chain, 141
Kuiper Belt, 116
Kurzweil, Ray, 26–27, 57, 61, 236

L

La La Land, 166
Lanier, Jaron, 22, 197
Laughlin, Gregory, 112
Lawnboy Ventures, 25
Lee, Pascal, 114
Lee, Richard, 43
Lenzi, Tommaso, 67
Levinson, Arthur, 51–52
Levy, David, 214–215, 217
Libella Gene Therapeutics, 56
Licina, Gabriel, 86–87
life extension, 51–59
Linghao, 176
Lithgow, Gordon, 53
Liu, Richard, 175–176
livestock
    antibiotic-resistant bacteria and, 72–73
    cloning, 62–63
    genetically modified low-fat, 71
    new animals, 72
    VR and, 183
Llewellyn, Dan, 128
Locus Biosciences, 72–73
Lomax, Beth, 118
longevity, 51–59
Loomis, Eric, 140
love
    computers and, 208
    robots and, 211–218

Love and Sex with Robots, 215
Lovetron9000, 43
Lowe, Derek, 47–51
Luckey, Palmer, 22
Lyme disease, 75

M

Ma, Jack, 226
machine learning. *See* artificial intelligence (AI)
Made In Space, 117
Magic Leap, 31
Makse, Keith, 173
malaria, 75
Malaysia, smart cities in, 136
mammophant, 129–130
managers, 218–220
Manin, Yuri, 96
Mann, Steve, 149
manufacturing. *See also* industry
    3D printing in, 176–177
    deep automation and, 174–175
    in space, 117
    market forces, control of, 225
Marrazzo, Jeff, 87
Mars. *See also* human expansionism; space
    colonizing, 111, 113–115
    journey to, 124, 125, 127
    landers on, 116
    living on, 131
    terraforming, 115
    water on, 119
Marx, Karl, 193
Mason, Christopher, 129
mass connectivity, xx. *See also* augmented reality (AR); brain-computer interfaces (BCIs); internet; virtual reality (VR)
    artificial sensory perception, 24–28
    combined with superintelligence, 260
    control of, 34–35
    local rules/norms and, 35
    machine-learning algorithms and, 2–3. *see also* artificial intelligence (AI)
    opting out of, 34–35
    reality and, 28–30, 33–35, 36
    seventh wave of, 35–36
materials, new, 99–102

INDEX

Mathis, Jeff, 56
Matrix, The, 27
Maxwell, Keith, 68
McGinn, Conor, 209
McLoughlin, Michael, 67
meaning, 193–194
meat
   cloned, 62–63, 64
   consumption of, 77
   genetically modified low-fat, 71
   lab-grown, 76–78
   mechanization. *See* automation; deep automation; robotics; robots
medicines, 47–51, 84, 87–88, 103. *See also* disease/illness; health care
memories
   generating, 14
   rewriting, 19
   transferring, 65, 256
Memphis Meats, 76–77
mental health, 148, 179–180, 207, 215–216
meta-learning, 189
Metformin, 56
Meyer, Michael, 119
mice, studies on, 57–58
Microsoft, 224
military. *See also* Defense, US Department of
   AI and, 233–234
   exoskeletons, 68
   gene drives and, 75–76
Min, Dongbin, 7
mind assistants, 15–16
mind hacking, 37
Minority Report, 140
Mitrokhin, Anton, 222
Miyako, Eijiro, 188
Miyashita, Homei, 26
Mojo Vision, 32
molecular programming, 78
Monsanto, 69, 70
Moon, 116, 117, 132
Moorehead, Scott, 25
Moore's law, 99
Mosa Meat, 78
mosquitoes, 75
movies, AI and, 171–172, 173
multivitamins, 50
Muotri, Alysson, 254

music, deep automation and, 167–171
Musk, Elon, 156
   on AI experts, 197–198
   colonizing Mars and, 115
   on dangers of AI, 230, 231
   Neuralink, 11, 20
   on simulations, 28
   SpaceX, 113, 114, 115, 116, 122–123, 124, 125

N

nanobots, 106–109
nanochips, 104
nanocondom, 103
nanoparticles, 102
nanoscale, 102–106
nanotechnology, xix, 95, 102–109
narrow AI (artificial narrow intelligence), 134, 196, 228–229
NASA, 114, 116, 118, 124, 132
National Federation of Retailers, 143
National Security Agency, 19
natural disasters, 183
navigation, 31
near-field communication (NFC) chips, 43, 44
Nectome, 237–238
Nejat, Goldie, 209–210
Nelson, Bill, 116
Nelson, Bob, 60
neocortex, synthetic, 27
Neom, 135
Nestor, Adrian, 7
Netflix, 3, 165
network effect, 158
Neurable, 5–6
neural implants, 8
Neuralink, 11, 20
neuroplasticity, 8, 249
neuro-suspension, 60–62
neurotech, 6. *See also* brain-computer interfaces (BCIs)
Neven's law, 99
New Zealand, 73–74
Newman, Stuart, 82
news, AI-authored, 172
NFC (near-field communication) chips, 43, 44

Ni Ziyuan, 147
Nicolelis, Miguel, 9–10, 239, 241, 245, 247, 252–253
noise pollution, 137
Nolan, Laura, 233
nootropics, 47–51
Norimaki Synthesizer, 26
nuclear war, threat of, 111–112
nursing homes, 185, 209–210
Nuyujukian, Paul, 11
Nvidia, 227

O

Oculus, 22
Omidyar, Pierre, 52
O'Neill, Gerard, 117
OpenAI, 224, 238
Oracle, 219
organ transplants, 80–83
organoids, 253–256
Orosei, Roberto, 119
outer space. *See* human expansionism; Mars; space

P

Pachet, François, 169
Page, Larry, 52, 156
paintings, by AI, 159
Pairwise, 70
pandemics, 112
panpsychism, 200–201
Parker, Kevin Kit, 77
parking, smart, 136
PASSAnT, 146
Paulus, Markus, 212
pavement, electrified, 136–137
Peng, Lily, 151
perception, augmenting, 247–253
Perry, Michael, 142
personal trainers, 210
personhood, legal, 204
Petomics, 40
pets, genetically altered, 85
Picasso, Pablo, 162
Pierce, Jessica, 64
piercings, 45
pigs, transgenic, 80–81

Pitzer, Bob, 181
placebo effect, 50
Plague Year, 109
Planetary Resources, 121
planets, life on, 128–132
plants, new, 40
plastic surgery, 46
poetry, by AI, 161
Poitevin, Helen, 219
Pokémon GO, 30
police, 140–143, 145–146
police violence, 149
Pollard, Barry, 62
pollinating, 188
pollution, 137
Poponak, Noah, 120
population, longevity and, 59
post-capitalist society, 194
posthumanism, 38, 131. *See also* biohacking
Pouratian, Nader, 25
power, 101, 137
Preceyes, 178
prediction, 149–154
PredPol, 140
Priebe, Sofia, 87
printing press, 1
privacy, 17–18, 144–145, 217
programming, 198–199
Project Debater, 154

Q

quadriplegics, 10–11, 67
quantified-self movement, 48
quantum computing, 96–99. *See also* human expansionism
quantum entanglement, 127–128
quantum internet, 128
quantum physics, 95, 100–101
quantum world, xx. *See also* human expansionism
quasiballistic photons, 13
Queisser, Tobias, 165

R

Rando, Tom, 55
Rapamycin, 55–56

reality
  brains and, 247
  control over, 257
  mass connectivity and, 36
  mixed, 35
  simulations indistinguishable from, 28–30
reality, mixed, 33–35
Rebrikov, Denis, 92
Red Meat Games, 173
relationships, 215–216
reproduction. *See also* sex
  artificial womb, 130
  designer babies, 90–92
  gene editing of unborn children, 89–90
  IVF, 65
  nanocondom, 103
  in outer space, 130
  preimplantation genetic screening, 90
  with robots, 216
  three-parent babies, 90–91
restaurants, robot, 190
retail, 175–176, 177
Richardson, Kathleen, 217
rights, of robots, 202, 204
Roberts, Todd, 19
Roberts, Tristan, 88
robotics, 134, 175. *See also* deep automation
robots. *See also* automation; deep automation
  attributing moral status to, 212
  BCIs connected to, 252
  diversity of, 203
  eldercare and, 185–186
  emotionally responsive, 207–211
  evolutionary, 222–225
  humanoid, 205–207
  love and, 211–218
  materials for, 188–189
  meta-learning by, 189
  origami, 177–178
  and possibility of jobless society, 192–194
  responsibility for, 204
  rights of, 202, 204
  self-learning, 220–225
  sex and, 213, 214, 215, 217
Rockefeller, John D., 225

Rocket Labs, 122
Rodriques, Samuel, 102
Rogue One, 171
Rosenberg, David, 184
Ross, Pablo, 81
Rossant, Janet, 80
Rothblatt, Martine, 236
Rothschild, Lynn, 118
Roy, Pierre, 168
rubber hand illusion, 247–248
Rudovic, Oggi, 207
Ruelens, Michiel, 166
rules/norms, local, 35
Rus, Daniela, 177–178
Russell, James, 74
Russell, Stuart, 198
Rutherford, Ernest, 198
Ryba, Nick, 26

S

Sabeti, Arram, 52
Sagers, Cynthia, 70
Sailor, Matt, 147
Sakamoto, Saki, 210
Samsung, 32
Sanchez, Steven, 68
Sanctuary AI, 205
Sankai, Yoshiyuki, 186
Santagate, John, 141
satellites, 122–123, 124
Saudi Arabia, 135
Schawbel, Dan, 219
Scheutz, Matthias, 214
Schmidt, Eric, 121
Scholl, Zackary, 161
schools, 155. *See also* education
screenplays, 166, 172
ScriptBook, 166, 172
Second Sight, 24
security. *See also* police
  automated, 141
  encryption, 98, 99
self, 237
self-awareness, 195, 199–202. *See also* consciousness; superintelligence
self-experimentation, 41–43, 51, 86–87, 88
self-identity, 246
self-improvement, 50

self-organized criticality, 237
self-worth, 193–194
Sen Gupta, Anirban, 107
senescent cells, 53
SENS Research Foundation, 61
senses, 24–28, 247–253
Sensorama, 22
sensors, 134, 137
sensory augmentation, 43
sentient machines, 202–205. *See also* consciousness; self-awareness
sex. *See also* reproduction
   biohacking and, 49
   Lovetron9000, 43
   reproduction and, 216
   robots and, 213, 214, 215, 217
sex slavery, 229
sexual transmutation, 49
Sharkey, Noel, 204
Sharman, Helen, 128–129
Shay, Jerry, 56
Shen, Solace, 187
ShotSpotter, 145–146
silicon, 129
Silver, David, 221
simulations, 27–30
Singapore, 137
singularity, economic, 191–194
Sinogene Biotechnology, 65
skills, soft, 156
slavery, 229
smartphones, 20, 215
smell, 25
Snow, Allison, 71
Snowden, Edward, 19
social intelligence, 207–211
social norms, 65
social order, deep automation and, 134
social programs, 192
Soderstrom, Tom, 124
SoftBank, 210
solar flares, 112
Solomon, Scott, 131
Sony, 167–168, 169
Sooam Bioengineering Research Institute, 64
SOSO, 7
sousveillance, 149

Southern, Taryn, 168
space, xx, 95. *See also* human expansionism; Mars; Moon; SpaceX
   advertisements in, 122
   asteroid mining, 120–122
   debate over colonizing, 132
   interplanetary ecosystem, 116–120
   life on other planets, 128–132
   manufacturing in, 117
   reproduction in, 130
   satellites, 122–123, 124
   space economy, 123–125
   terraforming, 115, 119
Space Angels, 121
space economy, 123–125
space elevator, 127
space habitats, 117–118
space industry, funding of, 123–124
space junk, 123
space tourism, 124
space-time, distorting, 126
SpaceWorks, 125
SpaceX, 113, 114, 115, 116, 122–123, 124, 125
Spark Therapeutics, 87–88
speech
   silent, 12
   turning brain signals into, 11
Spemann, Hans, 62
spinal anesthesia, 41
sports, 194
Spotify, 169
Squirrel AI, 155
Sriram, Sharath, 104
Stanley, Jay, 148
Star Trek, 44
Starkey, Natalie, 112
Starlink, 122
StartRocket, 122
statistics, 163
Steinberg, Jeffrey, 89–90
stem cells, 81, 82. *See also* gene editing/genetic engineering
stereoscope, 22
stereotyping, 144. *See also* biases in AI
Strauss, Karin, 79
Streisand, Barbra, 63, 65
subconscious, 241, 242

INDEX

subvocalization, 12
suicide, 151
suits, haptic, 23
super sentience, impact of on humanity, 256–257
superintelligence, xx, 134. *See also* consciousness; intelligence explosion; self-awareness
   AI economies, 225–228
   benevolent, 230, 232
   combined with mass connectivity, 260
   emotional intelligence, 196
   emotionally responsive robots, 207–211
   empathy and, 196
   love and, 211–218
   merging with, 235–238
   multiple, 204
   possibility of, 196
   rights and, 202
   robot bosses, 218–220
   safeguards for, 230–235
   self-learning robots, 220–225
   timing of, 197–199
supervolcanoes, 113
supplements, 47–51
supply chain, deep automation and, 176
surgery, 178–179
surveillance, 143–149
SweetPeach, 40
Szilard, Leo, 198

T

TALENS, 85
Tan, Anderson, 121, 122
Taoist formula, 49
tardigrades, 132
taste, 25–26
tattoos, 45
taxation, 192–193
teachers, 155. *See also* education
technology
   merging with biology. *see* bio convergence
   as part of system, 2–3
Tegmark, Max, 200
telepathy, spontaneous, 4
teleportation, 127
telomeres, 53, 56
ten Kate, Inge Loes, 119
terraforming, 115, 119
tests, standardized, 155, 158
therapist, virtual, 179–180
Thiel, Peter, 61
Thinking, Fast and Slow, 241
thoughts. *See also* brain-computer interfaces (BCIs)
   conscious, 240–241
   rearranging, 13
   transferring, 10
3D printing, 176–177
Thrun, Sebastian, 242
Tian Xue, 105
TikTok, 169
Timo-Iaria, César, 9
Tirumalai, Madhan, 131
Tononi, Giulio, 200, 201
transhumanism, 38. *See also* biohacking
transplants, 65–66, 80–83
trans-species, 248
Traywick, Aaron, 88
Truman Show, The, 149
Tumlinson, Rick, 121
Turing test, 236
Turkle, Sherry, 208, 215
Turned On: Science, Sex and Robots, 213
Twist, Markie, 214

U

Uber, 226
Ultra Safe Nuclear Technologies, 126
United Therapeutics, 80
Unity Biotech, 53
universal basic income, 192
university admission, 158
US Bionics, 68

V

van Gogh, Vincent, 162
van Wingerden, Peter, 183
Velsberg, Ott, 139
Venter, Craig, 80
Verdin, Eric, 53

Verver, Leon, 146
Vijg, Jan, 52
Vinu, Ajayan, 105
Virgin Galactic, 124
virtual experiences, 27–28
virtual reality (VR), 3, 21–24, 155, 183
vision, 24, 43, 87, 249–250
Vocaloid, 213
VR (virtual reality), 3, 21–24, 155, 183
VRX, 31

W

Walgamott, Keven, 67
Wang, Harris, 130
Wang, Xudong, 107
warping, 126
Warwick, Kevin, 42–43, 254
water, drinkable, 104
wearable devices, 163
Weihenmayer, Erik, 250
Welch, David, 77
Wells, H. G., 80
wetware, 41
Wheatstone, Charles, 22
White, Harold, 126
Wicab, 249–250
Winfield, Alan, 227
Winkle, Katie, 210
Wishnatzki, Gary, 181–182
womb, artificial, 130
Wong, Gerry, 174
Wong, Ian, 188
Wong, Shing-Chung, 104

Wordsworth, Robin, 119
work, end of, 191–194
workplace, robot bosses in, 218–220
World Health Organization, 151
World View, 124
writing, 1. *See also* art/design; entertainment
Wulf, Sylvia, 71

X

Xi Jinping, 174
Xiang Ye Tao, 136
Xu, Ming, 53

Y

Yamada, Kazuko, 210
Yan, Hao, 108
Yang, Andrew, 192
Yeltsin, Boris, 111
Yong Li, 85
YouTube, 3
Yun, Joon, 52
Yuste, Rafael, 236

Z

Zayner, Josiah, 85, 87, 88–89
Zero 2 Infinity, 124
Zhang, John, 90
Zhang, Xiang, 105
Zuckerberg, Mark, 13, 151

ABOUT THE AUTHOR

STEVEN HOFFMAN, or Captain Hoff as he's called in Silicon Valley, is the chairman and CEO of Founders Space (FoundersSpace.com), one of the world's leading incubators and accelerators. He's also an angel investor, limited partner at August Capital, serial entrepreneur, and author of several award-winning books. These include *Make Elephants Fly* (MakeElephantsFly.com), *Surviving a Startup* (SurvivingAStartup.com), and *The Five Forces* (TheFiveForces.com).

Photo by Naomi Kobuko

Hoffman was the founder and chairman of the Producers Guild Silicon Valley Chapter, served on the board of governors of the New Media Council, and was founding member of the Academy of Television's Interactive Media Group.

While in Hollywood, Hoffman worked as a TV development executive at Fries Entertainment, known for producing over a hundred TV shows, acquired by MGM. He went on to pioneer interactive television with his venture-funded start-up Spiderdance, which produced interactive TV shows with NBC, MTV, Turner, Warner Brothers, History Channel, Game Show Network, and others.

In Silicon Valley, Hoffman founded two more venture-backed start-ups, in the areas of games and entertainment, and worked as Mobile Studio

Head for Infospace, with such hit mobile games as *Tetris*, *Wheel of Fortune*, *Tomb Raider*, *Thief*, *Hitman*, *Skee-Ball*, and *X-Files*.

Hoffman went on to launch Founders Space, with the mission to educate and accelerate entrepreneurs. Founders Space has become one of the top start-up accelerators in the world. Hoffman has trained hundreds of start-up founders and corporate executives in the art of innovation and provided consulting to many of the world's largest corporations, including Qualcomm, Huawei, Bosch, Intel, Disney, Warner Brothers, NBC, Gulf Oil, Siemens, and Viacom.

Hoffman earned a bachelor's degree in computer engineering from the University of California and a master's degree in film and television from the University of Southern California. He currently resides in San Francisco but spends most of his time in the air, visiting start-ups, investors, and innovators all over the world.